Mentoring Scientists and Engineers

Mentoring Scientists and Engineers

The Essential Skills, Principles and Processes

John Arthurs

WITH CONTRIBUTIONS FROM

IAN M. GRAHAM, GUS HANCOCK, AND TIM BRUNDLE

CRC Press
Taylor & Francis Group
Boca Raton London New York

CRC Press is an imprint of the
Taylor & Francis Group, an **informa** business

A BALKEMA BOOK

CRC Press/Balkema is an imprint of the Taylor & Francis Group,
an informa business

© 2022 Taylor & Francis Group, London, UK

Typeset in Times New Roman
by codeMantra

Library of Congress Cataloging-in-Publication Data Applied for

Published by: CRC Press/Balkema
           Schipholweg 107C, 2316 XC Leiden, The Netherlands
           e-mail: Pub.NL@taylorandfrancis.com
           www.crcpress.com – www.taylorandfrancis.com

ISBN: 978-0-367-72398-9 (hbk)
ISBN: 978-0-367-72400-9 (pbk)
ISBN: 978-1-003-15464-8 (ebk)

DOI: 10.1201/9781003154648

Contents

Foreword

John Arthurs' book on mentoring scientists and engineers crosses some significant boundaries. It transfers the insights of modern coaching philosophy and psychology for practical use in mentoring within science, technology and engineering. The book fully achieves what it sets out to do in the Introduction – to provide guidance on skills, principles and processes in mentoring for those who are thrust into the role with no previous experience of being a mentor. In reading it I have been impressed by the clarity and flow of the writing along with stories, a sound structure and the careful selection of coaching/mentoring models. Short enough to read in a sitting or two, with lots of examples and guidance around skills that are easy to refer back to, it will appeal directly to its scientific and technical audience.

In the Academy of Executive Coaching, we have developed and brought coach training to many people representing a wide variety of backgrounds in many different countries over the last 20 years. In this we have noticed that relatively few of our trainees have backgrounds in the physical, biological and earth sciences or in the many different fields of engineering and technology. The world needs more people who are skilled at helping younger scientists and engineers to think independently, to inspire them to create and explore new ideas and communicate effectively, above all, to develop new patterns of ideas that encompass not just small parts but the whole world.

Civilisation advances because of the work of explorers and teachers. Explorers in the broadest sense are those who look for new ideas and learn how to use them, especially scientists and engineers. Teachers in the broadest sense are those who pass on ideas and skills to other people. It seems to me that mentors in science, engineering and technology are doubly valuable because they fill both roles at the same time. I hope that this book will make a significant contribution towards achieving that aim.

John Leary-Joyce

Author's Note

In the 1970s John Leary-Joyce was one of the United Kingdom's pioneers of Gestalt therapy and training. By integrating his clinical and business experience in the 1980s, he went on to create an innovative and very successful training service and Gestalt-based coaching practice. John founded the Academy of Executive Coaching in 1999, one of the world's most certified learning providers in this field with accreditation from the industry's leading professional bodies – Association for Coaching, European Mentoring and Coaching Council and International Coaching Federation. Over the last 20 years AoEC has trained more than 13,000 people worldwide. Headquartered in London, it now has 14 global locations with 90 professional trainers. As one of the most inspirational teachers I've ever had the privilege of meeting, John uses his dazzling skill to respond flexibly and creatively to each of his students individually.

JWA

Acknowledgements

Along with my grateful thanks and appreciation, I gladly acknowledge my personal debt to my own mentors who have helped me at many stages throughout a 53-year career. The most important early formative influences were geology professors Adrian Phillips, Chris Stillman and George Sevastopulo in Trinity College Dublin (1963–1967) and later Prof. Rex Davis at Royal School of Mines, Imperial College London (1979–1980). I have been very fortunate in having many colleagues, friends and managers who helped me in so many ways with essential professional development. Among most influential of these was Harry Wilson (1921–2004), Director of the Geological Survey of Northern Ireland and later Deputy Director of the British Geological Survey. In 2009 the addition of executive coaching to geoscience came late in my career. I am grateful to Coaching Development of London for their basic training in executive coaching in 2008–2009 and to the Academy of Executive Coaching (AoEC) for further training in 2014–2015. I am especially grateful to John Leary-Joyce, founder of AoEC and an internationally well-known coach, not only for his inspirational teaching and supervision but also for writing a Foreword to this book.

This work grew out of a manual written to supplement a series of geoscience mentor training workshops in 2015–2019 for one of the world's oldest and most prestigious learned societies, The Geological Society. In this, Bill Gaskarth, former Chartership Officer of The Geological Society, promoted the workshops and gave me much encouragement. Of the eclectic mix of ideas in this book, many were developed and applied to mentoring from those in the well-known and authoritative books on executive coaching, interpersonal skill development and intercultural studies listed in the

References. David Clutterbuck, John Leary-Joyce and the Chartered Institute of Personnel & Development kindly gave permission for direct quotes from their seminal works.

I am especially grateful to friends in other branches of science and technology, Ian Graham, Gus Hancock and Tim Brundle, who have contributed case histories. Paul Lyle, fellow earth scientist and writer, and Roy Gamble, retired senior civil servant and classicist, kindly read through the manuscript and contributed a great deal of useful comment. I am humbled by the skills of the contributors and readers, any one of whom could have written this book much better than I have. I owe a great debt of gratitude to my mentees who have taught me a so much over the years, in many cases more than I have taught them. The useful practical advice of Alistair Bright and Marjanne Bruin, editors from Taylor & Francis, was essential in getting this work to completion.

Finally, without the support and encouragement of Trudy Arthurs, my wife, this work would not have begun, let alone completed. As a well-known executive coach and coach trainer herself, Trudy helped me to become trained in coaching. She has contributed extensively to this project behind the scenes. I could not have had a better companion in this enterprise.

John Arthurs

Author

John Arthurs, a Chartered Geologist, specialises in mentoring, executive coaching and training to support vocational professionals, mainly geoscientists working internationally in the mining, engineering and environmental industries. In a career spanning 53 years John has travelled widely, often living in remote places and meeting people of many different cultures. He has worked with mining, prospecting and engineering companies, third-level education and government geological surveys. Until official retirement in 2002, John was Director of the Geological Survey of Northern Ireland, the government's chief advisor on earth science policy. Since then he has been consulting in the United Kingdom, Ireland, Ghana, Burkina Faso, Guyana, Romania, Tanzania, Mozambique and Zambia.

Chapter 1

Introduction

1.1 WHAT THIS BOOK IS ABOUT

Being pitched into the role of mentor without any prior training is a common experience for senior practitioners in the world of science, technology, engineering and mathematics (the STEM fields). For many otherwise highly trained experts who find themselves in this situation, mentoring can cause anxiety, frustration and stress. Most of us are following a wholly absorbing vocation. Up to this point we will probably neither have had the time nor the inclination to seek mentor training. If that describes your situation then 'Mentoring Scientists and Engineers' is expressly written for you. It offers an introduction to mentoring along with practical guidance on the basic skills, principles and processes in what is intended as a quick and easy read. The harder bit comes with the practice.

Being a new mentor describes my situation about 20 years ago, engaged as I was in a busy career in applied geoscience in mineral exploration and government scientific service. Now, looking back, I wish I had known the principles, process and skills covered here at a very much earlier stage in my life. They are things that I learned by receiving good mentoring and professional executive coaching, by undergoing coach training myself and then through years of practice. It's not that the difficult interpersonal situations I met on the way would not have arisen, they probably would. It's rather that I would have known better how to manage such situations and so created better outcomes. As an expression of my own learning, this book is an eclectic and idiosyncratic compilation of many different sources and subjects ranging through coaching and mentoring, popular psychology and social science, communication and management studies. The ideas are generic and widely applicable.

The book was written in the hope that they will help you in whatever branch of science, technology and engineering that you practise.

As professional scientists and engineers we are all the beneficiaries of a magnificent history of scholarship that stretches back over a thousand years. While our universities do an excellent job of primary training, think back to how you were when you graduated or how it is now to work with recent graduates. We see performance gaps showing up in unexpected ways. Although not lacking basic technical knowledge, it is clear that recent graduates do not have all of the skills they need. Common lapses appear as ignorance of essential techniques, in critical thinking, in communication skills both written and interpersonal, in under-developed professional attitudes and in understanding of our employer's purpose and organisational culture. All these kinds of knowledge and skills we lump together under the vague heading of 'experience'. When you have this experience we believe, then you become really useful and effective in your job. In spite of excellent post-graduate technical training and the wide availability of training courses in the so-called soft skills, the process of gathering professional experience is still mainly informal, *ad hoc* and unstructured. In fact, we learn mainly from our peers and our seniors on the job. That begs the question at the heart of this book – *what is the best way to help our younger colleagues gain experience and develop as professionals?*

There is a widespread belief, in my view mistaken, that mentoring is just a kind of on-the-job, one-to-one technical training that some people mistakenly call coaching. This kind of informal training is what is called in the UK Civil Service, 'Sitting with Nellie'. If you have either been Nellie or sat beside her, you will know that she needs much more than knowledge of the job alone to be effective. She must also be in possession of a specific set of interpersonal skills. You'll have an inkling of what these skills are if you reflect on good mentoring you have received in the past, perhaps mentoring that inspired you and opened your eyes to what you really wanted to do in your life and career. In later life it may have occurred to you to ask yourself, "How exactly did she do that?"

Looking back into our own personal history, everyone has felt the influence of other people on the development of their career. Learning from a mentor ranges on a spectrum from an occasional, casual and incremental piece of useful knowledge at one extreme, to a series of life-changing realisations that influence the whole of one's life and career at the other. As a mentor you can derive enormous satisfaction from watching your mentees' successes. It is, however, a vicarious satisfaction. One of the most powerful motivations to become a mentor is to repay the debt that you feel you owe for knowledge and skill that you received from your own mentors in the past.

Social science research over the last two decades has confirmed our everyday wisdom that says mentoring helps both mentees and mentors. Repeated surveys of successful mentoring among widely diverse organisations have shown that both mentees and mentors draw higher salaries, climb their career ladders faster and are more satisfied with their careers overall. See, for example, Allen et al. (2006) and Holland (2009) (*Note:* bibliographic details are given at the end of the book in References).

Managers of STEM organisations carry the responsibility to ensure best practice, to protect their organisations against the risk of professional error and to encourage their professional employees to develop their careers. While formal training is the proper response, in fact many conventional training courses yield a disappointing return on investment. Research studies have shown that mentoring in support of training courses greatly helps trainees to remember and to transfer the material they have been learning into their occupations. See, for example, Olivero et al. (1997).

Formal mentoring schemes may seem like the obvious management response to promote and enhance the very significant benefits that mentoring can provide. In spite of the best intentions of their organisers, however, these formal schemes are often dysfunctional and a great many just fade away. One of the UK's leading and best-known authorities on mentoring, David Clutterbuck, writes:

> Estimates of what proportion of mentoring programmes fail to deliver significant benefits vary widely, depending on how success or failure is assessed. A good working estimate, however, would be that at least 40% do not meet one or more of the following criteria:
>
> - Achieving a clear business purpose (e.g. improving retention in a target group of mentees by 25% or more)
> - Achievement of most mentees' personal development objectives
> - Learning by most of the mentors

- Willingness of both parties to engage in mentoring (as mentor or mentee) again... and many meet none of these.
  (Clutterbuck, 2011, with kind permission of the author)

So how should we avoid dysfunctional mentoring? Everyone knows there is more to it than just giving some good-natured advice. But what exactly is that 'more'? Since 2016, one of the world's great learned societies, the Geological Society, has sponsored training workshops for mentors, represented by groups of very experienced and highly qualified senior geoscientists. As the organisers of these events, we asked the participants before they started what it was that they wanted to learn about mentoring. At the time we were surprised by the high proportion of this group of very experienced scientists (97%) who seem to have found themselves thrown in at the deep end of mentoring without any training at all.

The overwhelming majority (92%) wanted to find out about interpersonal mentoring skills, processes and structure and not, as you might expect, tips on giving technical instruction in geology. Just over a third of the total sample (36%) simply wanted some generic guidance, such as, "An increased awareness of techniques to assist individuals in meeting their full potential", "To have a clearer understanding of the differences between mentoring and coaching" and "Where is the line (if any) between technical mentoring and personal mentoring?" In a related request, about 15% wanted to know how to structure the mentoring process, for example, "... strategy for mentoring, including how to prepare for meetings with mentees. Also, advice on what I need to do better/differently for the mentee to get the most out of my mentoring". For many, confidence in their own ability to mentor was the major issue. Echoing the views of many, one person asked, "What level of responsibility do I have for my mentee?" With a hint of desperation, another reflected, "... worried that I didn't know what I was doing." One remarked, "I find it difficult to discuss challenging behaviours and issues". These concerns were closely related to problems about motivating early career scientists who are resistant to gaining chartership and in other matters. Commonly desired interpersonal skills included managing the mentoring relationship and expectations around managing time, such as, "How much to give and how to cope with mentoring in the midst of an already busy work schedule?"

Overall, a feeling of floundering, worry and frustration ran through the responses to our questionnaires. It is now abundantly clear that a large majority in our sample of otherwise successful geoscientists found themselves in a mentoring role without really understanding what they should be doing. Since then, talking to colleagues in other scientific and engineering professions, it has become equally clear that this experience is widespread.

Mentoring is simple but not necessarily easy. No great effort is required to study the literature, but adherence to a small number of principles and exercising specific skills is absolutely necessary for mentoring to be effective. Many of the skills are closely similar to those practiced in counselling but applied in a non-therapeutic context. These skills rest firmly on a set of professional ethics and a client-centred approach. Some mentors have spontaneously developed these ideas and skills for themselves and so need little training. Most of us, however, need some guidance and practice to enhance our mentoring within our own professions. This book provides an introduction.

To begin with, all mentors need three essential attributes:

1 Specialist knowledge and skill in the same profession as their mentees,
2 Possession of, or willingness to acquire, the knowledge and interpersonal skills needed to create a productive mentoring relationship, and,
3 The time and inclination to be a mentor,

What you will find in subsequent chapters is:

Chapter 2. What is a Mentor? This chapter provides definitions, gives an overview of what a mentor does and shows how that differs from other learning and development techniques.

Chapter 3. Essential Mentoring Skills. A good mentor is not only a high performing practitioner in his /her own profession but also exercises five essential interpersonal skills: (1) Skilful questioning, (2) Active listening, (3) Building trust, (4) Self-management, and, (5) Giving constructive advice and structured feedback. The first four skills (questioning, listening, self-management and trust building) form part of the more extensive skill set of professional and executive coaching. The final skill (giving advice and feedback) is that of professional advisors such as lawyers and doctors. Finally, other desirable competencies are briefly described. They include explaining, raising awareness, designing strategies, using intuition and understanding leadership issues.

Chapter 4. Mentoring Principles & Process. Practical advice is given about how to manage the mentoring process and reduce the risk of a dysfunctional relationship. The meanings and consequences of each of five principles are explained: (1) Creating awareness is the primary purpose, (2) Client-centred mentoring, (3) Self-responsibility, (4) Intrinsic self-motivation and (5) Ethical responsibility. A four-stage structured framework for a mentoring engagement in a short series of time-limited sessions is described:

1   Connecting, how a mentoring pair comes together
2   Scoping explores whether or not the pair actually should work together. If agreed, a mutual understanding of what is to be achieved and how it is arrived at
3   Mentoring sessions opening with 'contracting', an informal but explicit agreement about expectations and intentions, and
4   Final review and closure maintains the focus on the specific outcomes of the series.

Chapter 5. Mentoring in Practice. Common mentoring applications and challenges that mentors encounter are considered. Establishing a baseline, understanding the mentee's knowledge level, thinking skill and learning style, is where mentoring should begin. Recommendations are made on how to carry out mentoring for technical knowledge and skill development. Helping mentees to gain professional qualifications, manage dysfunctional attitudes, make career transitions, and acquire self-confidence are discussed. The complexities of mentoring in intercultural circumstances are briefly considered.

Chapter 6. Mentor Training and Organisational Mentoring Schemes. Developing mentoring skill is best learned through observed practice in one of the training programmes accredited by the major coaching and mentoring institutions. Supervision, which is mentoring for mentors' practice improvement, is recommended. An organisational mentoring scheme is outlined. It requires the mentors to be trained, to adhere to mentoring principles described here and to structure the engagements. It is essentially voluntary and relies on confidentiality.

Case Histories accompany each chapter in order to give some real-world context to the text. Case History 1.1 is an account of a successful mentoring engagement that illustrates many of the points covered in the rest of the book. Many of the case histories elsewhere have been contributed by practitioners in other disciplines in order to give a different perspective from my own as a geoscientist. Their contributions are brilliantly illustrative and cleverly apt. All the contributors are distinguished people in their own fields of medicine (Ian Graham), chemistry (Gus Hancock) and technology (Timothy Brundle) (see below, Section 1.2).

Individual and interactive exercises are included because the majority of learning is experiential and comes through practice and self-reflection. This book started as a practical manual to accompany a series of training workshops for mentors. Its structure and layout are therefore suitable to accompany either a training course presented to a group or individual self-directed study.

References to bibliographic sources named in the footnotes are given at the end.

Figure 1.1 is a line drawing of a classical temple as a metaphor and an *aide-memoire* for the elements of mentoring. The whole structure is founded on principles. The elegant dome is the mentoring process. The five columns that support the dome represent the essential mentoring skills – active listening, skilful questioning, building trust, advice and feedback and self-management. The latter are described in Chapter 3 while principles and process are covered in Chapter 4. The first thing, however, is to think about what mentoring really is. That comes next.

1.2 INDIVIDUAL CONTRIBUTORS OF CASE HISTORIES

Throughout history the best way of communicating ideas has always been the telling of stories. In the present context, I can tell stories about my personal experience of mentoring in applied geoscience. What I cannot do is tell stories about how mentoring is practiced within other professions. Therefore, a number of senior practitioners in science, health care and technology have kindly provided examples of mentoring case histories from their own careers. All are distinguished experts who have made valuable contributions in their own fields. These case histories illustrate the principles and practice of mentoring as they are applied within each contributor's own professional experience. Taken collectively they provide essential illustrations of the generic process and principles

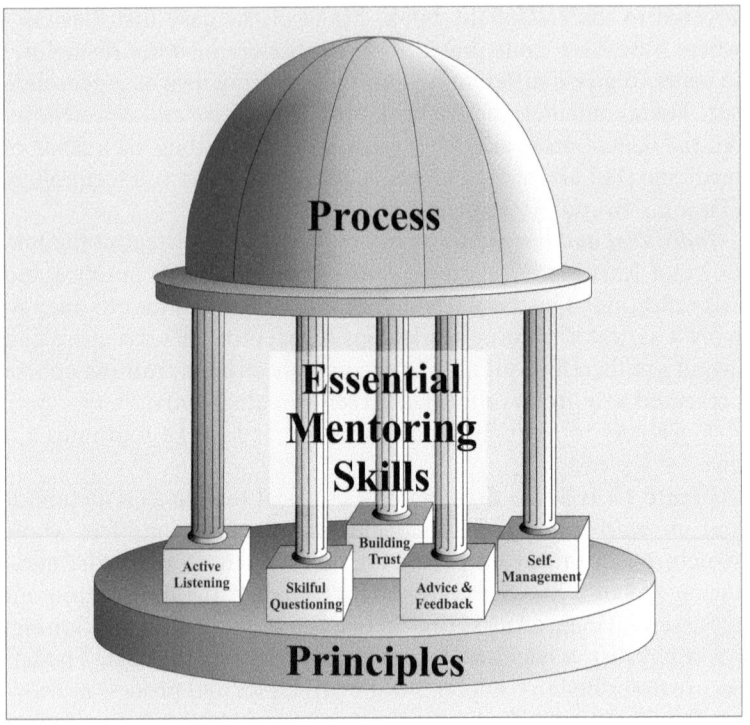

Figure 1.1 The elements of mentoring.

of mentoring. I should add that any names and circumstances of people mentioned in the examples have been changed to preserve their anonymity. The case histories appear in separate text boxes at the end of sections and chapters.

The individual contributors are:

Ian M. Graham – Professor of Cardiovascular Medicine, Fellow of Trinity College Dublin and Secretary to the European Society of Cardiologists. Now in late career, Ian is one of Ireland's leading and best-known cardiologists, not only at home but also more widely in Europe. He has travelled extensively, teaching cardiology to health care professionals and mentoring them in many countries. As a member of the Cardiovascular Round

Table Task Force, a European initiative, Ian's research interest includes prevention of cardiovascular disease and its communication challenges. Ian's vocation, wisdom, compassion and dedication have deservedly led him to the highest level of responsibility.

Gus Hancock – Professor of Chemistry and former Head of the Physical and Theoretical Chemistry Laboratory, University of Oxford, Emeritus Fellow of Trinity College Oxford. Gus is distinguished for his numerous definitive laser-based investigations that span broad areas of gas kinetics, energy transfer, reaction dynamics, photochemistry and medical diagnostics. He has won the Reaction Kinetics Award, the Chemical Dynamics Award, the Corday Morgan and Polanyi Medals of the Royal Society of Chemistry, and the 2000 Italgas Prize for Science and Technology for the Environment. In a world of very clever scientists and brilliant teachers, Gus is simply one of the cleverest minds, clearest communicators and kindest people I have ever known.

Timothy Brundle – Director of Research and Impact, Ulster University. Tim is responsible for the operational leadership of Ulster University's research strategy, governance and administration and guides its impact through knowledge transfer and intellectual property commercialisation. He is also Chief Executive of Innovation Ulster Ltd, Ulster University's award-winning venturing and investment company. Tim's deep understanding of research and innovation is motivated by his supportive interest in innovators as people and driven by his energy, enthusiasm and sense of fun.

CASE HISTORY 1.1 – A RECENTLY GRADUATED GEOLOGIST LEARNS TO THINK INDEPENDENTLY (JWA)

In 2004 a mining company retained me as a consultant to prepare a geological map in a remote area of eastern Zambia that is prospective for copper and gold deposits. Geological mapping is used in prospecting to interpret and understand geological theories that might be applied to locating mineral deposits in the area. Much more is involved than in topographic surveying. In addition to identifying rock outcrops and plotting their location, geological surveying requires the geologist to infer the nature and location of boundaries between rock formations unseen under soil cover and to develop conceptual models of the geological structure and of the physical and chemical processes that could create ore deposits in the area. In this case the geological models I was tasked with creating would be used to identify prospecting targets that would be drilled in due course.

The area was uninhabited woodland savannah country with various natural hazards including wild animals. The mining company provided two local hunters as guides and a recent graduate in geology as a field assistant. I was also required to train the latter, Emanuel, on the job. He was conscientious, paid attention and listened carefully. He was competent in identifying rocks and minerals in hand specimen and plotting outcrops. It took me a while to realise that something about our work together was not going well. If I asked a question that called for an inference from observations, he looked awkward and either parroted what I had said previously or waffled about some vaguely relevant theories. After a few days I privately began to doubt that Emanuel could have carried out a geological survey on his own.

Now, in all our interactions Emanuel was unfailingly polite and deferential. I knew that he would see me as 'Mudalla', a term of respect in the Bemba language for an old and supposedly wise person. One evening while sitting relaxed around the campfire after dinner I asked his permission to enquire about a sensitive topic. I assured him it would be confidential. When he agreed I said, "In my country we gauge a student's interest and progress from the questions he asks. I notice that you're not asking me any questions about our field work. Please can you tell me what is stopping you?" He looked away and shifted uncomfortably, apparently struggling to answer. I waited quietly. At some point in his own inner dialogue he must have decided to trust me. Slowly at first and then in a rush he told me that in his university, asking questions was impertinent or foolish. It would be seen as either a challenge to the lecturers' authority or as a demonstration of culpable ignorance. Students who asked questions were liable to be marked down. In my case, Emanuel simply saw himself as too junior to ask questions. At that moment he must have feared for his job and, indeed, for his whole future career. Clearly his admission took a great deal of courage, a willingness to trust me and a huge desire to learn.

Emanuel's admission came as a relief for both of us. It was clear what we had to do next. I explained that engaging in technical discussion was essential for his professional development. I reassured him that there are no stupid questions and showed him how to ask questions respectfully. After that there was no holding him. He quickly learned not only how to question me but also, and more importantly, how to interrogate the data we were collecting. Soon, I was able to send him off on surveying traverses by himself. On one such traverse he discovered the first outcrop of copper ore found during this phase of our investigation. In spite of Emanuel's success, the prospect did not prove to be economically feasible and the company eventually withdrew. Subsequently, he got sponsorship to attend an MSc course at a prestigious South African university from which he graduated with a distinction. When I last heard from him he was a regional manager for one of the world's largest mining companies, a rising star and a good mentor in his own right.

Emanuel's success is entirely his own. I can claim no credit. My intervention was a simple facilitation, no more than catalytic. Deconstructed in hindsight, it consisted of four simple actions:

1 Creating the conditions for trust and mutual respect to develop
2 Listening and noticing Emanuel's behaviour
3 Asking a simple but powerful question, the answer to which provided Emanuel with an insight
4 Discussing technical questions and encouraging Emanuel to be curious.

EXERCISE 1.1 – PREPARATION

We learn most efficiently and gain the most satisfaction from studying if we begin by identifying the subjects and techniques that we want to know about so that we can use them in specific circumstances. Then as we read, keeping a background awareness of our objective provides us with focus and the starting point for productive reflection and further learning.

- *Call to mind one or more younger professionals to whom you could offer support for their professional development.*
- *What do you know about their professional development needs?*
- *In what ways can you best provide them with professional support?*
- *What challenges concern you most in providing professional development support?*
- *What three things would you most like to learn by reading this book?*

What is a mentor?

2.1 DEFINITION OF MENTORING

Mentoring is one of the oldest forms of adult education, perhaps *the* oldest. From prehistory to the present day its most effective expression is story-telling. We have all heard myths and legends which include a character who offers knowledge and wisdom to the hero in a quest. The story-teller intends that we identify with the hero and learn the lessons that he or she learned. Among Western cultures some of the best-known examples of legendary mentors are Pallas Athene, Merlin, the Fairy Godmother, Gandalf and Obi-Wan Kenobi. The rich and diverse cultures of Africa and Asia similarly have their own characters fulfilling exactly the same mentoring role in their legends and stories. The Mentor of Legend is, in fact, an archetype for someone having wisdom, experience, knowledge and good judgement. While these are also the attributes of a good mentor in science, technology and engineering, for most of us it is a counsel of perfection, an ideal to aim for.

The dictionary definition of mentor is "An experienced and trusted adviser" (Concise Oxford Dictionary, 1964). In Greek mythology the hero Odysseus charged the original Mentor with training his son, Telemachus, in his absence fighting the Trojan War. Most of us know what we mean by the words 'mentor' and 'mentoring', but beyond the basic definition, do the words mean the same thing to all of us? In fact, these terms have acquired a variety of shades of meaning over the centuries. What may be included under the heading of mentoring seems to have expanded to overlap with the whole range of learning and development activities. Apart, perhaps, from the distinctive character and function of the Mentor of Legend, there is no widespread agreement about what mentoring is and what it is not. For the present purpose of helping mentors to improve their practice, it will help to define the term and then make some distinctions with other, related practices.

The world's best-known professional institution for human resources, the Chartered Institute of Personnel & Development (CIPD), defines mentoring, as follows:

> Mentoring in the workplace tends to describe a relationship in which a more experienced colleague shares their greater knowledge to support the development of an inexperienced member of staff. It calls on the skills of questioning, listening, clarifying and reframing that are also associated with coaching. One key distinction is that mentoring relationships tend to be longer term than coaching arrangements.
>
> (CIPD, 2020; with kind permission of the publisher,
> the Chartered Institute of Personnel and Development,
> London; www.cipd.co.uk)

The modern use of the word 'coaching', referring to executive and professional coaching, sometimes causes confusion. The distinctions between coaching, mentoring and counselling are made in Section 2.3 (below).

In practice, a mentor can take on some or all of four general functions:

1   *Technical mentor* – When we want to know some specialist scientific, technical or engineering knowledge or skill needed for our professional practice we may approach a senior colleague who is an expert in that specialisation. So, for example, senior surgeons are called upon to train colleagues in the practical details of particular procedures in which they are skilled. Engineering and scientific companies often pay academic specialists to train their employees in their area of technical expertise. Effective technical mentoring is more than just telling people things they need to know about the work. The mentor also needs instructional skills and is therefore also a teacher or instructor. Experienced and skilled mentors, however, avoid lecturing and information dumping. Rather they use the key non-directive skills of questioning, listening, clarifying and reframing, as referred to in the CIPD definition. The mentoring pair may go on to broaden the range of discussions

from technical instruction and in so doing develop a longer-term relationship covering other types of mentoring.

2 *Career mentor* – When we arrive at a decision point in our careers we may approach a senior colleague for help. This is usually someone with whom we have established a professional connection and whose technical knowledge and experience we have learned to trust. We may be looking for detailed information and advice, perhaps discussion about strategy for advancement or performance feedback or just general guidance about working in various organisations, institutions and sectors. In this situation a mentor cannot fully know what is in another person's best interest and so it is essential to take a cautious non-directive approach (see Sections 3.5 and 3.6).

3 *Support mentor* – Nobody gets through a career without meeting some difficulties on the way. By virtue of longer personal experience mentors can and do provide personal and emotional support. In fact, such support is at the heart of collegiality. It is underpinned by professional friendship, acceptance, acknowledgement and ultimately trust (see Section 3.4). The CIPD definition refers to the similarity between mentoring and coaching. Various authors have described the full spectrum of different types of coaching, including performance, development, skills and leadership varieties. To this list, Leary-Joyce (2016) adds existential coaching which addresses deep issues of personal identity and purpose at the other end of the spectrum. It is important for mentors to recognise that all these types of coaching and mentoring involve sensitive areas. It is easy to make damaging mistakes and fracture a relationship. In some cases of this kind it may be more effective to direct the mentee to professional coaching by a specialist in the relevant area.

4 *Role model* – Because the mentor is an expert in the same field as the mentee, he or she quite naturally provides a role model for the mentee to follow. When faced with a situation which has been previously discussed the mentee will naturally think, "I should do here what my mentor would do in this situation". Prolonged mentoring instils in the mentee the same professional values as those of the mentor. Ideally, mentoring should be inspiring and provide a vision for the future. Enthusiasm is always the best teacher. Tim Brundle's story (see Case History 2.1) is an inspiring example of how role modelling works in practice.

Some typical mentoring challenges and applications that belong in one or other of these general functions are considered in Chapter 5.

In discussing how mentoring and developmental relationships work generally, Murphy and Kram (2014), go into much more detail about the variety of mentoring roles. They recommend that rather than having a single mentor, a mentee should create a *'developmental network'*. That is, an informal group of people who are prepared to offer help to individuals. Employers can help by providing the means and support for creating such networks.

CASE HISTORY 2.1 – MESSAGE FROM A ROLE MODEL – TOMORROW BELONGS TO THOSE WHO CAN HEAR IT COMING (TIMOTHY BRUNDLE)

'Tomorrow belongs to those who can hear it coming', was the slogan David Bowie coined to promote *Heroes*. It neatly captures one of his most important talents: to sense the future and draw it forward into the popular culture of the present. Sometimes he would grasp the importance of a trend, but more often it was his artistry in self-reinvention that opened up new modes of cultural expression or brought deeper social change soaring up to the surface. A good mentor enables self-reinvention.

Growing up in Northern Ireland, I wanted to live in the future, not in the past. I became fascinated by the relationships between people and technology, began to recognise technology as a means of human realisation, and understood the critical role of research and experimentation as a foundation of societal progress.

I studied science, philosophy and economics and to keep me in university, I took a job carrying the bag of a man who has been asked by the government of the day to set up a science park in Northern Ireland, which is today known as Catalyst. His name was Professor Ernie Shannon and he was my friend and first mentor. He taught me many things, but most important was that innovation is about people and very little to do with technology.

Professor Ernest Shannon, 'Ernie', who died in 2011 aged 73, was an engineer, inventor and scientific advisor to government. Born in North Belfast, educated through an apprenticeship and later at Queen's University of Belfast, he guided a team of 350 British Gas engineers in the development of 'pipeline investigation gauges' or 'PIGS'. The 1970s had exposed weaknesses in gas pipelines across the world that had resulted in fractures, explosions and considerable environmental damage. The PIGs zoomed about Britain's 10,000 miles of pipeline to collect data for Ernie's engineering team to detect potential weak spots. For producers of the James Bond films, meanwhile, pipeline pigs emerged as useful and suitably technological plot devices, enabling 007 and his accomplices to cross borders undetected.

To get people in the room and to get people talking, Ernie encouraged me to arrange Northern Ireland's first innovation conference. I wrote to a bunch of people asking if they would speak in the hope that the presence of someone interested might encourage people to turn out. The author Douglas Adams said yes.

Douglas stepped off the plane in Belfast from a visit to Apple where he had been spending time with Jonny Ive and Steve Jobs. He declared that Apple would soon be one of the most important companies in the world. He explained Jonny's design philosophy, which was to care deeply about the people who were engaging with the technology, simplify the experience around their needs and aspirations, and to solve every problem in a way that acknowledges its context.

In his conclusion, Douglas told the cautionary tale of the Great Irish Elk. The Irish Elk, *Megaloceros*, is misnamed, for it is neither exclusively Irish nor is it an elk. It is a giant extinct deer, the largest deer species ever, that stood over 2m at the shoulder, with antlers spanning up to 4m. The Irish elk evolved during the glacial periods of the last million years. It ranged throughout Europe, northern Asia and Africa. The name 'Irish' stuck because well-preserved fossils of the giant deer are especially common in the peat bogs of Ireland. It was thought that the Irish elk finally went extinct when the antlers became so large that the animals could no longer mate. His conclusion being that the elk had started evolving on an irreversible trajectory towards larger and larger antlers, which ultimately resulted in extinction. He urged attendees to set aside past trajectories for the world of technology would soon get in the way. To think differently, just like Apple, to re-invent ourselves.

Ernie and I spent long hours drinking tea in a muddy portacabin in Belfast's docks trying to figure out how to aggregate and align the interests of those in the region who were active in R&D and who might back a plan to build one of Europe's largest science parks. Over tea, he taught me three lessons. The first was that innovation doesn't occur in a vacuum. Innovation is not something you can do to someone but rather with someone. Second was that innovation didn't stick or sustain in environments without a culture of experimentation and where people feared the risk of failure. The third was the importance of universities, upon whose research new technologies were forged and with whose skills value from technology was created. The science park was a success, as were many of Ernie's endeavours. But more than the brick and steel, I like to think that his values and ideas are embedded in the foundations of Northern Ireland's knowledge-based recovery.

2.2 ORGANISATIONAL LEADERS AND MANAGERS

So, in trying to establish what mentoring is, it is actually easier to say what mentoring is not. Where do we draw the line?

The most basic duty required of a leader is to make plans, find resources and issue instructions intended to protect and develop the organisation for which they are responsible. The word 'instruction' has the dual meaning both of giving orders and of teaching about something. Those receiving the orders must know not only what to do but also how to do whatever it is they are called on to do. In practice managers operate by a variable combination of 'command-and-control', consultation and teaching depending on who they are managing at the time. So where is the line to be drawn between managing and mentoring?

One much quoted definition of mentoring is:

Off-line help by one person to another in making significant transitions in knowledge, work or thinking.

(Megginson *et al.*, 2006)

Here the term 'off-line' refers to someone who does not have direct management responsibility for mentee. This definition emphasises the distinction between management and mentoring functions. Managers have a primary duty to the organisation they lead, whereas a mentor's primary duty is to their individual mentees. Conflating the two roles can give rise to a conflict of interest. For example, how can a manager who is charged with implementing redundancies also fulfil his duty of care for his mentees whose jobs are to be retrenched at the same time?

Mentoring is a personal and professional development activity intended to help the individual career of a younger professional. As an essential aspect of their professional practice, all scientists and engineers must take responsibility for their own actions and opinions. Therefore, while a manager has a duty to give orders, in contrast, a mentor should not order or direct a mentee to do something because that would cut across the mentee's acceptance of their own professional responsibility. Modern coaching and mentoring are therefore ***essentially non-directive and facilitative.*** (See Section 4.1).

In what has been called sponsorship mentoring, a common variation is that of an older and more experienced person who supports and promotes a *protégé* (*French*: one who is protected). Although there is nothing wrong in principle with giving active practical help in advancing a mentee's career, care has to be taken that ethical problems do not arise. A company chairman, for example, who provides introductions and opportunities not given to other employees at the same level risks accusations of favouritism. If that chairman then covertly seeks some personal advantage in promoting his mentee, say by gaining a lucrative consultancy contract after retirement, then an ethical boundary has been crossed. The mentoring relationship involves not only a duty of care but also one of ethical responsibility.

2.3 EDUCATIONAL PROFESSIONALS AND MENTORS

We can make an essential distinction between, on the one hand, school children in secondary education and, on the other, adult learners in third-level education and in continuous professional development. For children we expect school teachers to take responsibility for training their pupils (technically *pedagogy*). In contrast, we expect adult learners to take responsibility for their own development (technically *andragogy*). This means that teaching adults can really only be successful for self-starters who want to learn. It also means that professional adult learners, in particular, must learn the essential skill of independent thinking. That is best imparted by role modelling and a non-directive mentoring style (see Section 3.5).

Those of us who left school and went directly on to undergraduate training in university experienced the transition from juvenile to adult learning. At university we had tutors and lecturers who were employed to deliver structured courses of information which we were supposed to augment by private study. From those who were good teachers we learned the basic information, the thinking patterns and practices that are characteristic of our different professions. Most significantly we learned to study by and for ourselves. Some tutors and study supervisors may have been sufficiently helpful with personal and career development matters to be regarded as mentors in the true sense. In my own case, for example, one of my lecturers helped me get my first professional job, something that gave him no benefit personally and was not a contractual obligation of his employment. It was quite simply motivated by a concern for my benefit which was also true for other students at that time.

Education for the adult learner therefore must be a voluntary activity. Mentoring is more concerned about helping than directing. You cannot coerce someone into being mentored. In fact, self-motivation is the *sine qua non* for professionalism. Adult learners, however, do not necessarily know what they need to learn in advance and therefore some guidance is usually necessary. In formal adult education we can make a distinction between lecturers who deliver technical information through lectures and tutors (or study supervisors) who give advice on what to study. Good tutors are recognised by the effectiveness of their discussion and guidance and by a proper concern for their individual students. In this way good tutors could be described as effective mentors. The subtle difference between academic tutors and mentors in the wider world is to be found in the essential voluntary nature of mentoring. Effective engagement in mentoring can be only by mutual agreement. In contrast, few (if any) tutors can decline to work with the tutees given to them by the university admissions procedure.

Professional institutions and learned societies all require that post-graduate practitioners ensure their own best practice and protect their employers and clients against error as a condition for maintaining their professional status. Best practice is achieved therefore by and through self-motivated continuous professional development (CPD). Formal training schemes to help their employees with their professional development are now very common within the larger scientific, engineering and technical firms. Given titles such as *Training Supervisor, Tutor, or Study Director*, the managers of such schemes are usually senior practitioners within the organisation. These training supervisors are required to advise their trainees on the subject knowledge required for proficiency in a specialist area in addition to their substantive professional responsibilities for their employer. As with university tutors, a training scheme supervisor can also provide mentoring.

If you have been put into the role of Training Supervisor you might worry about how much instruction or training you have to deliver, what is the best way to carry it out and whether or not you have all the necessary knowledge and skill. Remember that your role as a Training Supervisor is to guide the trainee, not necessarily to do the instructing yourself. If the scheme also requires you to do some instruction, that is an additional role. If you are faced with a mentee who resists training, the real question is not how they should be trained, but rather what is stopping them from learning for themselves? Why they are not appropriately motivated? The subject is covered below (see Section 5.5).

Many inexperienced mentors think that their mentees are best served by simply giving them as much information as they can think of. Saying things like, "Here's a list of 80 journal articles and five textbooks you must read for this subject", or delivering a 20 minute *impromptu* lecture on something the mentee has casually asked about simply leads to overwhelm. Often born out of a desire to give the greatest possible help, 'data dumping' is one of commonest mistakes that inexperienced mentors and tutors make. If we reflect on our own experience, we already know that helping intelligent, self-motivated people to find out something for themselves is always more effective than spoon-feeding. Modern mentoring is, as far as possible, non-directive.

Rather than simply providing information, you can best help your mentees by setting learning tasks and following up with questioning, discussion and argument. In this case a mentor performs much in

Table 2.1 Some essential differences between instructors and tutors

Instructor, trainer, lecturer	Tutor, study supervisor
• Formal training is structured around a pre-defined field of knowledge	• Formal duty is to direct the tutee to the required topics that he/she needs to study. May include structured tutorials in small groups
• Decides what trainees need to know within that field and delivers appropriate instruction	• Teaching component is usually unstructured, occasional and *ad hoc* according to the tutee's needs in the moment
• Conventionally delivers training as a series of structured lessons or lectures giving information in a systematic progression	• Guides or suggests where to find information for himself/ herself. Offers discussion and argument
• Lessons may include exercises and discussion	• Helps the tutee to work out solutions for themselves
• A less personal relationship, delivered to groups, both large and small	• A more personal relationship, delivered individually or in small groups

the same way as does a tutor in our traditional universities. This subject is discussed further below (see Section 3.4). The main distinctions between instructors and tutors are summarised in Table 2.1.

2.4 PROFESSIONAL AND EXECUTIVE COACHES, COUNSELLORS AND THERAPISTS

Simply grouped together as 'coaching', professional and executive coaching is a modern learning and development practice in business and the professions. It is nowadays practised by qualified practitioners for whom it is a profession in its own right and most are independent practitioners brought into organisations on a short-term contract basis. Some larger organisations employ internal coaches who fulfil the same function except that their practice is wholly for the benefit of the staff of the organisation they serve. With the advent of coaching modules in business and management schools over the last decade or more, a trend for leaders to use coaching as a management tool has arisen. Confusingly, the

"Sitting with Nelly" variety of informal training for technical and procedural matters is also traditionally called 'coaching'. However, a clear distinction should be made between that kind of informal instruction and modern professional and executive coaching.

The Chartered Institute of Personnel & Development describes coaching at work as follows:

> Coaching aims to produce optimal performance and improvement at work. It focuses on specific skills and goals, although it may also have an impact on an individual's personal attributes such as social interaction or confidence. The process typically lasts for a defined period of time or forms the basis of an on-going management style. Although there's a lack of agreement among coaching professionals about precise definitions, there are some generally agreed characteristics of coaching in organisations:
>
> • It is essentially a non-directive form of development.
> • It focuses on improving performance and developing an individual.
> • Personal factors may be included but the emphasis is on performance at work.
> • Coaching activities have both organisational and individual goals.
> • It provides people with the opportunity to better assess their strengths as well as their development areas.
> • It is a skilled activity, which should be delivered by people who are trained to do so. This can be line managers and others trained in coaching skills.
>
> (CIPD, 2020; with the permission of the publisher, the Chartered Institute of Personnel and Development, London; www.cipd.co.uk).

The modern practice of mentoring is developed from professional and executive coaching and therefore uses similar models and skills. *Mentoring is a specialised application of executive coaching in which the mentor is also experienced in the field that the mentee wishes to study.* Both mentoring and coaching are disciplines that share a distinctive set of principles based on a client-centred focus, openness, self-responsibility, trust and confidentiality and techniques. Mentoring and coaching are compared and contrasted in Table 2.2.

Table 2.2 Some essential differences between mentors and executive coaches

Mentors	Professional and executive coaches
• Mentoring is provided by an experienced professional in the same field as the mentee	• Executive coaches are <u>not</u> subject experts in the same field as the client
• Mentoring is not the mentor's primary occupation	• Coaching is a professional practice in its own right
• Non-directive in personal and career matters but can be directive in technical matters when a non-directive approach is not possible	• Wholly non-directive and limited to non-technical matters
• Aims to instruct &/or advise in technical fields. Non-directive as far as possible	• Aims to elicit, explore and facilitate client's own thinking about the career and/or personal problems presented using non-directive approach
• Relationship is typically longer lasting and not necessarily restricted in content or time	• Relationship is typically restricted to a limited series of sessions to work on one issue over a number of weeks or few months
• Good practice requires clear mutually agreed structure like that of executive coaching	• Professional coaches are required to structure sessions in order to achieve the agreed outcome
• Mentors always in the same or related profession as the mentee and usually in the same organisation	• Executive coaches are usually external, brought in by agreement to work with specific individuals

Although the principles and techniques used in professional coaching originate from counselling and therapy, it is important to make a clear distinction. A ***counsellor*** is a professional advisor who specialises in helping to resolve personal and emotional problems. A ***therapist*** (also psychotherapist) treats mental disorders by psychological or talking therapy rather than by medical intervention. ***Mentors and coaches are not trained to treat mental or emotional disorders and should never attempt to do so. In fact, they have a duty of care to refer their mentees to appropriate specialists should the need arise.***

2.5 WHEN MENTORING GOES WRONG

By understanding where things can go wrong we can try to avoid mistakes and improve our own practice. The old saying, "There's no such thing as a bad student, only a poor teacher", is obviously simplistic but it does carry a grain of truth. Since experience and position power is on the side of the mentor rather than the mentee, in case of difficulties the first place we should for solutions is in the mentoring practice.

Although most mentoring relationships are successful to a greater or lesser degree, some do end in failure. By 'failure' we mean that they did not achieve their intended purpose. Clutterbuck (2011) estimates that 40% of mentoring schemes fail. While some of that failure is undoubtedly due to administrative factors, such as lack of organisational support and inadequate provision of time for mentoring, the remainder is due to failure of individual mentoring relationships.

Eby *et al.* (2000) carried out a survey of mentees' negative experiences of mentoring within two large executive development schemes. Out of a total of 240 mentees 84 reported at least one negative individual relationship. The total of negative experiences at 35% is not too far off Clutterbuck's 40% estimate for failure of mentoring schemes (Clutterbuck, 2011). Analysis of mentees' accounts of 168 negative experiences allowed the identification of five groups of negative behaviours exhibited by mentors. The group providing the most frequent source of problems is mismatching of values, work styles and personalities (27%). Distancing behaviour, including neglect, self-absorption and intentional exclusion, is next most common source (24%). The manipulative behaviour group includes misusing power of position, tyranny, sabotage, credit-taking and politicking. Incompetency in either interpersonal or technical matters is grouped together as a lack of mentor expertise. The fifth group, general dysfunctionality, includes attitudinal and personal problems. Overall, the behaviours shown by the mentors concerned are strikingly inappropriate. If nothing else, these results clearly illustrate why mentor training and voluntary mutual agreement in mentoring pairs are essential considerations.

Less serious mentoring problems are very common as I can attest from making my own mistakes as well as observing those of colleagues and of mentors in training. Individually, a single mistake may not amount to much and the relationship may not be permanently damaged. However, repetition and accumulation of these errors will inevitably impair mentoring to the point where the relationship breaks down. Some of the more common mistakes are due to:

- *Poor communication* is the cause of most problems. Making assumptions about what other people know and don't know leads to miscommunication. Similarly, vagueness and its opposite, data dumping, gives false impressions of precision, understanding or completeness.
- *Assumed equivalence* includes offering guidance based on the mentor's own experiences. Saying, "If I were you I would…" is invariably unhelpful and can be disastrously misguided. All individuals are unique with differing personal histories.
- *Complacency* can cause the mentor to assume that the solution to the mentee's problems is already clear. It can lead to overlooking unique factors and a failure to investigate all the issues fully.
- *Pride* can cause the mentor to assume that he/she knows more than they actually do. It is essential to have the relevant knowledge and appropriate experience. If not, then it is important to be honest about the limits of one's knowledge and experience.
- *Cognitive biases and emotional blind spots* are, by their very nature, unconscious but they are always present and everyone has a unique set. Dobelli (2013) describes no fewer than 99 cognitive biases, others say there are many more. Such biases are compounded by a lack of self-awareness and imprecision of thought. In the worst cases the mentor's set of unconscious biases can be imprinted on the mentee.

Mentoring is a self-evidently two-sided process. Obviously a mentee's behaviour can also be dysfunctional. However, if the contracting procedure is correctly carried out and permission to hold the mentee to account is agreed (see Section 4.2), then the mentor has a responsibility to give the mentee honest feedback on their dysfunctional behaviour (See Section 3.6).

Mentors can avoid stumbling into dysfunctional relationships by:

1 **Developing and practicing the basic mentoring skills (see Chapter 3),**
2 **Adhering to the mentoring principles (see Section 4.1), and,**
3 **Structuring the mentoring process (see Section 4.2).**

EXERCISE 2.1 – REFLECT ON YOUR OWN MENTORS

List as many of those people who have been involved in your personal and professional development at different stages of your life as you can think of – not only mentors and coaches, but also parents, relatives, teachers, advisors, instructors, *etc.*

Focus on two people who had a beneficial influence:

PERSON A – Name or other identification

- *What did he or she do and why was it so influential?*
- *How was that learning, knowledge or skill transmitted and why was the approach so effective?*
- *What did you learn about effective mentoring from that experience?*

PERSON B – Name or other identification

- *What did he or she do for you and why was that so influential?*
- *How was that learning, knowledge or skill transmitted and why was the method or approach so effective?*
- *What did you learn about effective mentoring from that experience?*

Think of one person who tried to influence or instruct you but failed or was otherwise ineffective.

Note: the objective here is to understand mentoring mistakes and so avoid making them yourself.

Do not dwell on this part of the exercise if it makes you unduly upset or angry.

PERSON C – Name or other identification

- *What did he or she do and what was the effect on you?*
- *What prevented his/her actions from being effective?*
- *What did you learn to avoid in your own mentoring?*
- *What have you learned about ineffective mentoring you have received in your own career?*

Chapter 3

Essential mentoring skills

3.1 MENTORING AS AN INTERPERSONAL COMMUNICATION SKILL

Academics, learned societies and professional associations all have well-established views of what it means to be competent within their own professions. As a generalisation, the wide variety of knowledge, skills and practice in science, engineering and technology can be divided into two groups of competencies – technical and professional:

1 *Technical competencies* are those that include the character-istic knowledge, thinking skills and practice of the technical, scientific and/or engineering material of the subject. That is everything we learned from textbooks, lectures, tutori-als, practical classes, laboratories, workshops, clinics and field trips in university and in continuous professional de-velopment (CPD) after that. This category could be further sub-divided into:
    - Cognitive skills – acquiring and using the knowledge base of the profession. For example, medical practitioners understand human anatomy, chemists know how elements interact with each other, geologists re-construct local earth history, civil en-gineers understand design principles for structures, *etc.*
    - Practical and manipulative skills – for example, paramedics know how to treat injuries, analytical chemists use laboratory instruments, architects visualise future buildings, surveyors draw maps, software engineers use coding languages, *etc.*
2 *Professional competencies* are those that are necessary for the effective practice of the technical competencies. That includes

effective interaction with clients, co-workers and all who have a stake in the work and may include specific aspects of other overlapping fields, *e.g.* specialised law, health, safety and environmental management practice. Ethics and codes of conduct are an essential part of all professions. In the present context of mentoring, the most important professional competency is communication skill. That itself covers a wide range of skills, including writing, presentation and interpersonal (or social) skills. The precise nature of the communication skills acquired by individual practitioners depends on their aptitudes and chosen career path.

Good mentoring is a professional competency in its own right. It is the optimum combination of the mentor's own technical competency with a specific set of interpersonal communication skills for the benefit of a colleague's development in the same profession.

The major coaching and mentoring professional associations include the Association for Coaching (AC, 2020), the International Coach Federation (ICF, 2020) and the European Mentoring & Coaching Council (EMCC, 2020). In addition to promoting coaching and mentoring, they also establish standards for practice, accredit training organisations and regulate CPD. Addresses for their websites can be found in the References. As in professional associations for all other professions, the coaching and mentoring institutions require their members to be able to demonstrate a specified set of competencies. In an ideal world, every mentor working in science, technology and engineering would receive training in mentoring skills and practices to at least Certified Coach level in the United Kingdom and Ireland or the equivalent in other countries.

The coaching institutions list a range of competencies all of which are generally useful in mentoring. Some of these competencies are actually necessary. However, coaches are trained not to give advice, whereas mentors must occasionally do so. Mentoring skill is therefore a combination of coaching skills together with advisory and instructional skills. While this covers a very broad range, for mentors who are approaching this subject for the first time, we suggest that *five interpersonal skills are essential:*

1   Skilful questioning
2   Active listening
3   Establishing trust

4   Self-management
5   Giving constructive advice and structured feedback.

For the sake of clarity these skills are described in separate sections below, although it should be apparent that they are mutually interdependent and must be employed interactively. So, for example, questioning and listening are two sides of the same skill set. Similarly, to have your advice listened to you must first have gained trust and be able to manage the session and your own behaviour appropriately. In general, the unique way in which individuals employ mentoring skills determines their personal style and how their presence as mentors comes across to their mentees.

3.2 SKILFUL QUESTIONING

The most obvious function of a question is to get information. Thinking about what other functions it can serve leads us to enquire into less immediately obvious aspects – its context, its underlying purpose and for whom the answer is really intended. At the heart of all mentoring is the way in which the mentor asks questions and listens. In a mentoring context it is clear that the more skilful the questioning, the better are the outcomes. In fact, the insight that a question draws out in the respondent is a measure of how good a question it is.

Powerful questions are the high points of any mentoring session. They are those that trigger a meaningful insight and lead mentees to change the direction of their thinking. In the right context the following could be powerful questions:

- "What's stopping you?"
- "What's most important just now?"
- "How do you feel about the solution you've generated?"
- "What might the real answer look like?"
- "If I were in your position, what would you tell me?"

Effective questions in mentoring have three characteristics:

1   *Simplicity* – Asking a short and simple question helps the mentee to frame a clear answer. When people ask long and complex questions, they either want to look clever or have started talking before being clear about what their real question actually is.

2   *Clarity* – Getting clear about the purpose of a question makes it more likely to elicit a useful answer.
3   *Open mind* – Asking a question with an open mind, detached but curious, makes it more likely that the response will be equally open and helpful.

At the opening of any social interaction, what psychologists call 'set induction', questions can and usually do start the conversation, *e.g.* "How are you?" Here, the opening question has the important function of setting the tone and nature of the interaction that is to follow. As a mentor, your opening questions and their accompanying non-verbal communication (NVC) carry important underlying messages about how you intend to interact with your mentee. If the sole purpose of a mentoring session is to help the mentee's professional development, then your questions should be for the benefit of the mentee and not for your own information or satisfaction. Appropriate humour is often a great start. Although it has an important purpose, mentoring does not have to be deadly serious. Intellectual playfulness of the kind shown by the Nobel-winning physicist, Richard Feynman, is a powerful and effective way of interacting with scientists, engineers and technologists. At the rational level of thinking (the cognitive level) your questions can arouse interest, develop knowledge and encourage critical thinking. When directed at the personal or feeling level (what psychologists call affect), your questions may unconsciously elicit emotional states and thereby lead your mentees to explore their own attitudes. Appropriately phrased questions show interest and suggest support and empathy that is essential for building trust and thereby achieving an effective outcome (see Section 3.4). Knowledge of different *types of questions* and their semantic functions enhances your flexibility in finding the right question to ask in a critical moment.

In the context of interpersonal communication skills, social scientists have studied questions and questioning in depth. Among the best-known authorities in this field are Hargie and Dickson (2004) along with other authors in the compilation by Hargie (2006). The following is a brief summary of the subject with particular relevance to mentoring.

Closed and open questions differ in the degree of freedom they allow for the response. Closed questions limit the response to binary answers, typically 'yes' or 'no' (*e.g.* "Is this a nitrogenous compound?") or ask for a simple identification (*e.g.* "What type of sample is this?"). Closed questions are generally unhelpful in mentoring because they constrain the mentee's answers and can cut off ideas. Open questions, by contrast, call for more information and so promote discussion. They are signalled by interrogative words. For example, in a technical discussion about surveying, notice the different responses that could be brought forth by varying the choice of interrogative words:

What seeks information and analysis, *e.g.* "What is Simpson's Rule?"

Where seeks location within space, process or argument, *e.g.* "Where could Simpson's Rule be applied in this project?"

How explores process and strategies, *e.g.* "How would you apply Simpson's Rule in this case?"

Which asks for a choice to be made between alternatives, *e.g.* "In which of these cases could Simpson's Rule be used?"

Who seeks a name or role, usually with more information, *e.g.* "Who first applied Simpson's Rule to calculate volumes?"

When seeks a time, usually with process or historical context, *e.g.* "When was Simpson's Rule used in this design?"

Why calls for reasons, causes or justifications, *e.g.* "Why would you use Simpson's Rule here?" In mentoring try to avoid using 'Why'. It can sound demanding or heard as criticism and can elicit a defensive response. Imagine how a young surveyor might feel if the Project Manager asked him the 'Why' question above in a sneering tone.

Cognitive and affective questions draw upon different parts of the mentee's thinking systems. Cognitive questions address the rational part of our thinking. In mentoring, they are used to draw out information about substantive processes, facts, theories, *etc.* (*e.g.* "What symptoms indicate that the patient might be suffering from coronary artery disease?"). Affective questions, on the other hand, address our emotions. In mentoring they ask how the mentee feels about the subject under discussion. The latter are especially useful in career and personal support mentoring (*e.g.* "How do you feel about asking for a transfer?"). Affective questions can also be important in technical mentoring when you suspect that something unspoken is blocking the mentee's understanding. Such a blockage will usually be found to have a negative emotional loading such as fear, shame, dislike, associated with a frightening, shameful or distasteful event in the past. The mentee can only make progress in technical understanding and get over the blockage if he/she brings the emotion to conscious awareness and commits to dealing with it. Affective questions are particularly important because they tend to promote rapport in mentoring (see Section 3.5).

Leading questions are posed specifically to elicit a response that the questioner wants to hear. Anyone who watches courtroom dramas on TV knows that barristers think they already know the answers to the questions they ask in cross-examination. Leading questions are often used by inexperienced mentors as a way of spoon-feeding information. By contrast, one of the most important aims in mentoring adults is to encourage them to think independently. If, by asking leading questions, you covertly tell your mentees what to think, then you risk distorting their thinking processes and ultimately, their own intellectual independence. Therefore, *avoid asking leading questions in mentoring.* This is especially important in career and psychosocial mentoring where the only satisfactory outcomes are those which the mentees derive for themselves.

Probing questions ask your mentee to expand on or to explore their initial response. They lead to a questioning sequence that is a normal part of the mentoring dialogue. Some varieties and styles of probing questions are discussed in more detail in the next section (see Section 3.3). Note that all probing questions have to be used sensitively. Depending on the context of the question, the tone of voice and the personalities involved, one mentee might view a sequence of probing questions as an interesting and collaborative discussion, whereas another might see it as a threatening challenge.

A questioning sequence, as the name suggests, is a sequence of question-and-answer following a particular line of thought. In mentoring it is essential that mentees do not perceive questioning sequences as a form of hostile interrogation, especially if trust has not first been established. It is essential that the mentee feels that he/she has been heard and their point of view understood. *Therefore, as an essential part of a questioning sequence, the mentor must use the parallel skills of listening, clarifying and reflective listening* (see Section 3.3).

The best-known variety of questioning sequence, *the funnel sequence*, begins with an open question, progressively followed by more focussed questions to arrive at a specific conclusion. Conversely, an inverted funnel sequence starts with a closed question about some detail and gradually broadens out to general principles. Funnel questions are commonly used in recruitment interviews in which the recruiter wants to investigate a generalised or vague statement made previously by the interviewee. The funnel questioning technique is also used in counselling and executive coaching sessions which are focussed on creating a practical outcome. It is, therefore, also an effective approach in most mentoring situations. As a fictional example, imagine a discussion between two engineers:

MENTOR: "What would you like to talk about today?"
MENTEE: "I'd like to know your opinion about the conclusions in Mitchell's paper on design principles that you gave me last week. How should we apply those to this project here?"
MENTOR: "Yes, that's interesting. I'd be glad to. Which part of Mitchell's work particularly concerns you?", and so on, delving into more detail.

Note that, despite the mentee's request for an opinion, the mentor did not immediately launch into a detailed account of Mitchell's paper and how it should be applied in this case. Instead, the mentor answered the mentee's question with a counter-question.

EXERCISE 3.1 – QUESTIONS ABOUT QUESTIONS

Think about your own experiences of skilful questioning:

- *Reflect on a powerful question that elicited an insight; one that either someone else asked you or you asked someone else. What was it and what happened subsequently?*
- *Think of a powerful question from a scene in literature or film. What was the title of the film and the context of the question? What part did it play in bringing the story forward?*
- *How did the answer change the direction of the plot?*
- *What made the question powerful?*
- *Why is asking a good question <u>always</u> more valuable to your mentee than giving a lot of information?*

3.3 ACTIVE LISTENING

Many people think that mentoring is just about telling their mentees the things that they think they need to know, or what used to be called 'coaching'. Modern mentoring takes a different approach. To understand why, recall the last lecture you attended. How much of it do you remember? Experimental psychologists have consistently replicated the classic study by Ebbinghaus in 1880 which showed that our rate of forgetting is a logarithmic decay curve. Only two hours afterwards, we will have forgotten at least 50% of what we heard. If the lecture was yesterday, it is very unlikely that we will be able to remember any more than 20% and probably only about 10%. Having a load of information dumped on us, however well-intentioned it may be, is quite simply an inefficient way of transmitting information.

Mentoring is first and foremost a listening art. It has been said that most people listen with the intention to reply rather than to understand. Your job as a mentor is to facilitate your mentee's thinking and not to try and do it for them, no matter how tempting that may be. If, during a mentoring session, you realise you're doing most of the talking, then I'd respectfully suggest that you just stop, sit back and listen. A good part of the mentee's learning process happens when he/she thinks out loud. The rest of it happens by reflection and practice outside mentoring sessions. Therefore, your mentee should be doing most of the talking. The mentee's contribution in a career or psychosocial mentoring session can occupy more than 90% of the total time. The remaining 10% is set induction, review and closing with a few well-chosen questions in between. Listening actively and empathically helps mentees to gain insight and to express themselves more effectively. For your mentee to have a sense of being heard and of feeling acknowledged, it is essential that you listen well (see below, Section 3.5). Case History 2.1 illustrates the beneficial effect of just sitting back and listening.

Hearing is a physical activity. Listening is a learned skill. In fact, listening is one of the first cognitive skills that infants learn (Hargie & Dickson, 2004). Later in life we take it so much for granted that the idea of training people to listen seems strange. In fact, thinking about how we listen helps to improve our listening skill. If we could somehow gauge both the strength of the mentor's <u>in</u>tention to understand and his/her level of <u>at</u>tention, then we would see a broad spectrum of listening skills, illustrated

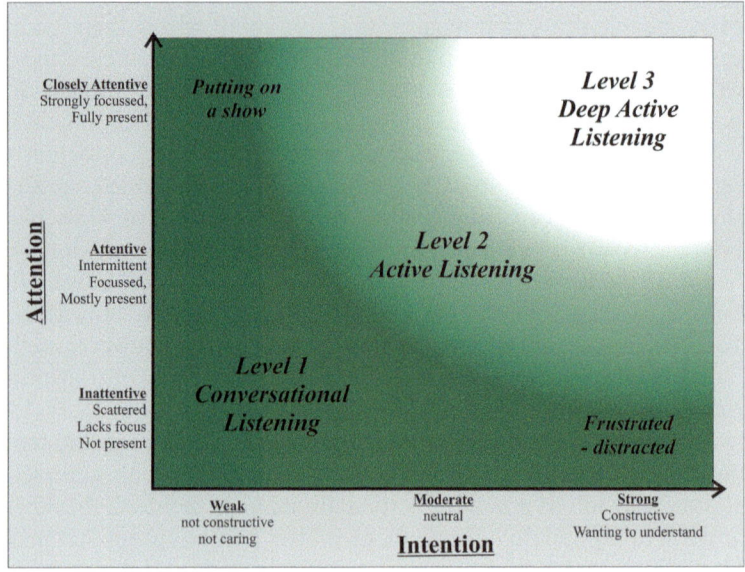

Figure 3.1 Three levels of listening.

diagrammatically in Figure 3.1. How to recognise and switch between three levels of listening during mentoring sessions is described in well-known textbooks on coaching, such as those by Starr (2003) and Hawkins and Smith (2006), and taught in accredited coach training programmes.

Level 1 – Conversational listening. For most of us, a conversational level of listening is a matter of everyday experience. In normal social circumstances our attention may be scattered and we may be distracted. Think, for example, of being at a gathering of acquaintances and strangers, say a reception or a drinks party. Although most of us are interested in other people to a limited degree, self-consciousness tends to rule in the moment. We are polite but pay only partial attention. We may feel we need to make a favourable impression. We hear our conversation partners' words but our own internal dialogue claims most of our attention. While they are speaking we are thinking about what to say next or perhaps whom to speak to next. The phenomenon of 'conference eye' refers to the frequently observed gaze behaviour in such settings. The speaker addresses the listener's name badge while the listener's gaze wanders off round the room. Conversational listening is of little use in mentoring.

Level 2 – Active listening. The people who stand out most clearly in your memory after the reception or drinks party will be the people who seemed to listen best. Although it is rarely taught as a skill in its own right, active or empathic listening is an essential skill in many professions, especially those that are concerned with supporting other people at a personal level, such as in health care, education and law. Many of those who are successful in professions that provide advisory services for clients, such as engineering, architecture and marketing, are also good listeners. In listening actively we pay close attention while consciously holding the intention to understand exactly what the speaker is saying. Our gaze behaviour and body language are consistent with paying attention. We unconsciously make small sounds (*ah-ha, mmm*), and changes in facial expressions and gestures that tell the speaker we are listening (nodding, small movements in rhythm with the conversation). We are mentally recording, we can playback accurately and we can ask clarifying questions and summarise. Active listening is the normal state required for a mentoring session.

Level 3 – Deep active listening. With practice and experience active listening can be brought to a deeper and even more focussed state. Most people will have had some experience of it, although it is not very common. Highly skilled and experienced counsellors, psychotherapists, executive coaches and spiritual directors exhibit deep active listening at least some of the time during their client sessions. To enter a deep active listening state during a mentoring session, the mentor holds a keenly focussed intention to understand all the spoken and unspoken aspects of what the mentee is presenting. He/she pays very close attention and shows empathy; all senses are alert; the mind is quiet and clear with no intruding thoughts; totally present and in the moment. The mentor hears, sees and senses not just what the mentee is saying, not only in words but also the message that their NVC carries. Intuitively, the meaning behind the spoken words becomes clear so that the mentor can respond to deeper issues. The mentee feels that they have been heard and understood and so enters their own optimum thinking state.

Dual processing is the skill of paying attention to more than one thing at a time. When reflecting on the differences between conversational and higher levels of listening, trainee mentors often ask, "How do I pay attention to what my mentee is saying and to my own thoughts at the same time?" To begin with, it requires awareness of one's own current level of skill, then with conscious practice and feedback the skill eventually enters our unconscious mental processing systems. In fact, dual processing develops quite naturally and unremarked in many other areas of life. For example, we learn to drive a car in four stages:

1   Unconscious incompetence – As small children, we are not aware of a desire to drive
2   Conscious incompetence – As young adults, we feel a need to learn but don't know how.
3   Conscious competence – As learner drivers, we know how to drive but feel awkward.
4   Unconscious competence – As experienced drivers, we drive the car and competently carry out other tasks at the same time, such as holding a conversation with passengers.

Reflective listening can be regarded as statements in the interviewer's own words that encapsulate and represent the essence of the interviewee's previous message (Hargie & Dickson, 2004). A lack of reflection in a mentoring conversation will incline the mentee to think that his/her point of view has not been understood. ***Reflection is an essential part of mentoring.*** In this context, the verb 'to reflect' has two meanings – both thinking about a memory and repeating some or all of what you've heard. A mentor's reflection focusses attention on particular aspects of what the mentee has said. It may check the mentor's perceptions or explore underlying meanings. By acting within his/her level of understanding, the reflection focusses attention on the mentee's own processes. As the large volume of research on interviewing skills has shown, when properly done reflection leaves the mentee with the feeling of being understood (Hargie, 2006). It is one of the main ways in which empathy is demonstrated; a subject covered in more detail below (see Section 3.5). Reflection differs from questioning, although their purpose can be similar. Used in the sense of responding, reflection is indirect and exerts less control on the tone and pace of the dialogue. By contrast, a question is more direct and includes more of the questioner's own life experience. In offering reflection it is important to remember that all our perceptions pass though the filters of our own experience and so our reflections are never without some distortion.

Reflective listening skills include the following varieties:

Echoing, or reflection *sensu stricto,* is simply repeating the mentee's own words. This method was developed to high degree by Carl Rogers (1902–1987), the originator of client-centred counselling, *e.g.* Mentee: "I don't understand why the company has treated me in this way". Mentor: "You don't understand why the company has treated you in this way?" In this case, echoing would draw out deeper thoughts and feelings about perceived mistreatment. In that it works very explicitly within the mentee's own frame of reference, echoing acts to suggest that the mentor accepts the mentee. A possible detraction is that it can cause annoyance and distraction if it is used continuously.

Paraphrasing, or reflection of content, restates what the mentee has said but in the mentor's own words. As such, it draws attention to what the mentor thinks that the mentee means. Paraphrasing offers the mentee an opportunity for revision and correction and so acts as a powerful way to help the mentee clarify his/her own thoughts.

Summarising is an even more concise statement than paraphrasing. After the mentee has made a lengthy or rambling speech, the mentor can summarise in just a few words. Again, this is a clarifying opportunity in that it encourages the mentee to focus on the main point of what they wanted to say.

Reflection of feelings infers the mentee's feelings from what he/she has said, and especially by observing non-verbal communication. It is used as a check on feeling and acts to draw out the underlying emotional content, *e.g.* "You sound hesitant. Perhaps you feel uncertain about that choice?" When sensitively done, and especially if used along with paraphrasing of content, this kind of reflection can lead to a powerful question, *e.g.* "I understand that you set up this investigation for the client's benefit, but he didn't accept the results. Can I take it from your tone that you feel annoyed?"

As a form of dialogue, mentoring consists of a series of questions and responses which flow from one to the next in a related sequence. For the mentor, questioning and listening are inextricably interlinked. Secondary questions follow naturally from the mentor's initial question or reflection. How they are phrased depends on what the purpose and intended outcome of the discussion is. Some variations of *secondary questions to aid listening* are:

Extending the discussion – Given a basic answer to the initial question, the mentor can encourage the mentee to expand on what he/ she has already said, perhaps with deeper analysis, *e.g.* "That's an interesting point you raise about the results of the strength tests. Can you say more about the likely consequences?"

Clarifying – is necessary when either you are confused or suspect that the mentee might be confused, *e.g.* "When you say ..., I'm not sure whether you mean this ... or that ...?"

Checking assumptions – when you suspect that an unrecognised assumption materially affects the mentee's conclusions and position, *e.g.* "In saying ...., what assumptions are you making?"

Checking alternatives – to see if the mentee's proposed alternatives really are mutually exclusive. Given a 'sucker's choice' of two equally undesirable alternatives, explore others.

e.g. Mentee: "I have two choices. Either I must do this or I have to do that..., which should I choose?" Mentor: "What other choices are possible?"

Exploring possibilities – Encourage your mentee to investigate the total range of possibilities which could exist; *e.g.* "What if...?"

Challenge generalisations – ask for specific examples; *e.g.* "When you say, 'These results are rubbish', do you really mean each and every single one? Which ones might be useful?"

Silence is golden, as the old proverb has it. A mentee's sense of being listened to is an essential aspect of all good mentoring. In using your active listening skills, NVC may be more important than anything else. Along with a calm posture and quiet but attentive gaze behaviour, few things will communicate your attention to your mentee more than saying nothing at strategic moments. In social situations, human beings tend to find difficulty with silence. We seem to have a need to fill the interactive space with speaking. If your mentee does not immediately respond to your question, do <u>not</u> give in to your natural tendency to fill the space by speaking; do not jump in with another question; do not try to help with suggestions and advice. Instead, just sit back and allow your mentee time and mental space to think through their response. A delayed response is often a precursor to a 'lightbulb' moment in the mentoring session. When such occasions occur, watch carefully. Your interruption of their thought process could blot out a significant emerging insight.

The thought process of scanning though memories, past experiences and half-remembered ideas is called *transderivational search*, a term derived from hypnotherapy. It is an inner search in the mind's eye maybe to find a reference experience that will help us to make some kind of sense of what could be a powerful question. The way you phrase your question can constrain or liberate your mentee's inner search. The process is usually marked by an unconscious but distinctive pattern of body language. The thinker adopts an alert posture and goes quiet. His/her gaze shifts away from the questioner to a faraway look in which the eyes move from side to side as if the mind's eye is watching an inner screen. The ideal outcome is a 'light-bulb' or 'eureka!' moment. The moment of insight is also often marked by a distinctive pattern of body language. The thinker relaxes, brings his/her gaze back to the questioner, smiles with an expression of satisfaction and says something like, "That's it!" Such wonderful 'lightbulb moments' mean that not only has the mentee achieved some kind of resolution, but also for you, as a mentor, it is a vicarious reward and a sense of satisfaction for a job well done.

CASE HISTORY 3.1 – A FRUSTRATED EXECUTIVE COACH LEARNS TO LISTEN (JWA)

Back in the 1980s before many of the techniques from counselling became normal practice in coaching, I worked with an executive coach, who I'll call James (not his real name). He was a well-educated Englishman recently retired from a multi-national company where he had enjoyed a high-achieving career as a senior executive. Over lunch at his London club James told me about one of his clients with whom he had experienced considerable difficulty. James' client continually ignored his well-meant advice and repeatedly failed to honour agreements made in previous coaching sessions. The client attended coaching sessions but was passively uncooperative. Eventually, James ran out of ideas and didn't know what to do next. In despair, he accidently solved the problem when he just sat and listened. In James' own wry comment, "I just gave him a jolly good listening to". With that the client opened up and began to respond more positively. Gradually it became clear that the client's line manager had sent James's client to him without any explanation let alone agreement. The manager was unhappy with the client's attitude and decided to get him some coaching in order to, "Sort him out". The manager and the client had a difficult working relationship and, as a result, the client mistrusted James from the outset. The client had a prior expectation that as his coach, James would tell him how to do his job, one that he felt he already knew how to do. However, James' simple act of listening unblocked the problem by itself.

Back in those days executive coaches were generally much more directive than they are now. They simply told clients exactly what to do. Nowadays, an executive coach would not accept an assignment such as this without first conducting a scoping session with the client and then a three-way session between himself, the client and the line manager. In this way the underlying problem would have come to light. Even if it was somehow missed, the subsequent contracting phase of the coaching dialogue would have revealed the situation and what could be done about it, if anything. Apart from the importance of listening well itself, a secondary point is that the mentor's active listening skill is exercised within a structured mentoring context. The process and structure of mentoring are discussed below (see Chapter 4).

3.4 AN ABILITY TO BUILD TRUST

It is a common experience that negative emotions such as fear, distrust and hostility inhibit learning. In this sense negative emotions range in intensity from mild annoyance and distraction to intense dislike which results in a complete learning blockage. The requirement for absence of negative emotion does not mean that effective mentoring necessarily has to be based on mutual admiration and personal friendship, although that can happen. The opposite of distrust is trust, which is essential for learning. Liking and friendship are optional extras. *Mentoring is a professional advisory relationship based on trust.* In legal jargon, it is a *fiduciary relationship* which means that the mentor has a duty of care towards the mentee. The best role models for this kind of relationship are experienced, wise and trusted family doctors and solicitors.

To develop and maintain your mentee's trust there are some essential do's and don'ts:

- Do your best for your mentee in all circumstances
- Never do anything intentionally that will cause gratuitous hurt or damage
- Maintain strong ethical and professional boundaries
- Maintain confidentiality and discretion about your mentee's information and concerns
- Be clear about what you can and can't do
- Stick closely to all agreements you make, however minor
- Develop your capacity for empathy
- Carry out "contracting" procedure – *i.e.* an explicit agreement on how mentoring is to work (see Section 4.3)

Trust is based on empathy, the compassionate understanding of the feelings of another person. It should be distinguished from ***sympathy*** which is an emotion, experiencing similar feelings, also feelings of concern and care for another person and wishing them well. In contrast, ***empathy*** is a cognitive ability, that of being able to recognize and understand the feelings of another person. Many people assume that empathy is something that we either have or don't have. Whereas in fact, it is a conscious response which can be enhanced by observation, reflection and training.

Rapport is the quality of a harmonious relationship in which two people empathise with each other's feelings, understand their ideas and communicate easily. They feel comfortable in each other's company, are able to disagree and free to act naturally. Rapport between any pair of individuals lies in a spectrum ranging from strongly positive feelings of attraction through to neutral. At the other end of the scale beyond rapport lie negative feelings of aversion. It is easier to create rapport if we are members of the same group (nationality, profession and sports team), speak the same language (word choice and technical jargon) and hold the same beliefs and values (politics, culture and aesthetic tastes). However, it is not necessary to share all these qualities to find rapport; it's just easier. Goleman (2006) gives a detailed account of how rapport works in practice.

Matching is a behavioural phenomenon between two people that indicates rapport. For example, we can observe matching when two friends who are enjoying each other's company sit together in a restaurant. As they gain rapport they unconsciously mirror each other's non-verbal language and body language, including similar pace of speaking, tone of voice, posture and gestures. *Mismatching* is the opposite of matching; that is when the pair are not in rapport. Mismatching is associated with feelings of social discomfort, anxiety, aversion, *etc*. If the couple in the restaurant are mismatching we don't need to hear what they're saying to each other in order to know they're having an argument. We just know by their actions – looking away, or looking intently with face contorted, arms folded, speaking at different rates, intensity and pitch.

Incongruence is evident when a person's NVC contrasts with what they are actually saying. Being unconscious behaviour, NVC in this context is graphically referred to as 'leakage'. To get a sense of what your own non-verbal leakage feels like, imagine receiving an unsuitable gift from a close relative whom you love and don't want to offend. How do you react? Most of us will smile and express thanks, but our incongruence will be evident from our hesitation, gaze behaviour and awkward gestures. *Congruence* is the opposite. Most humans and primates seem to have the unconscious ability to recognise matching/mismatching and congruent/incongruent behaviour.

As a way to develop trust, an effective mentor will aim to establish rapport with the mentee. It is important that you strive to remain congruent or you risk losing your mentee's trust. Looking for matching/mismatching and congruent/incongruent behaviour is a most useful practical skill that allows you to guide the mentoring session unobtrusively. Case History 3.2 illustrates how this works in practice. Exercise 3.2 describes how to practice gaining rapport.

CASE HISTORY 3.2 – A RECENT GRADUATE LOOKS UNCOMFORTABLE (JWA)

The context is mentoring a recent graduate on a topic within the field of structural geology. The mentor observes significant NVC during the following conversation:

MENTOR: When I asked if you had plotted a stereographic projection of the fracture planes, you said that you really hadn't had time yet. But I also noticed that you shifted in your seat, frowned and looked away. I'm guessing that you feel awkward about not completing the task and perhaps also you feel uncomfortable with this whole topic. Is that right?

MENTEE: [hesitantly] Well … honestly … at college I was ill for a while and missed all the classes on stereographic projections. I tried to read about it afterwards but there were a lot of other things going on at the time and I didn't really understand it. I admit that I feel awkward and uncomfortable with this topic.

MENTOR: OK. Thank you for telling me that. I think that stereographic projections will be an essential tool for understanding the controlling structure here. Would you like to learn more?

MENTEE: Yes, but I will need you to take it slowly.

MENTOR: OK, let's go into the field and take some fracture measurements. I'll plot one or two projections while you look on. Then you can plot some while I watch and we can talk about it as we go along. [They proceed as described. After a while …]

MENTOR: Do you think you're getting the hang of it now?

MENTEE: [Hesitantly] Well … yes … sort of.

MENTOR: Would you be willing to have a go at plotting all the points for this rock face and bringing the stereographic plots back here next Friday? I recommend that you first read Holcombe's article on his website. It's the most accessible description I know and it'll help you if you get stuck. That should answer most of the questions that might come up for you before Friday. Then we can have another discussion about how to use projections in mapping this fracture zone generally.

[MENTEE DEPARTS LOOKING MORE CONFIDENT]

Positive and negative language, and their relative frequencies can have very significant consequences on the development of trust in mentoring. Experimental psychologists can predict with a high degree of precision whether or not married couples will get divorced and business teams will succeed or fail (as described in Seligman, 2011). In observing meetings and interactions between married couples and within business teams, the researchers counted the number of positive interactions (encouragement, approval, affection, agreement, humour, *etc.*) and compared them with the total number of negative interactions (interrupting, harsh tones, abuse, aggression, *etc.*). In what has become known as the Losada Ratio or P/N score, repeated observations have found that the dividing line between success and failure is marked by a ratio of approximately 3:1 positive to negative interactions. Scores of less than 3:1 predicted that married couples are very likely to be divorced and that business teams will fail. On the other hand, with P/N ratios greater than 5:1, couples and teams flourish and individuals are happier. Unexpectedly, with a P/N ratio of greater than 11:1, teams begin to fail again. It seems that we require some honest feedback and moderate adversity in order to flourish. The Losada Ratio seems to have much wider applicability. In particular, the P/N number has important practical implications for mentoring.

As mentors we need to be consciously aware of the overall positive or negative tenor of our interaction with our mentees. If we are constantly criticising or pointing out difficulties, failures and problems with little or no acknowledgement to counterbalance, then our mentoring efforts will almost certainly fail. Alternatively, if we are too 'nice', with frequent compliments and much encouragement but never offering any challenge, then we will also fail. Although we may both enjoy our meetings, we will find that our mentee doesn't make much progress beyond a certain point. It follows that we also have to pay attention to how our mentees express their own patterns of positivity and negativity. If your mentee is self-critical or lacks confidence, then you can help by giving constructive feedback. Help them to see their real strengths more clearly, challenge their own internal negative thought patterns and encourage them to practice more positive self-talk.

Challenging and holding your mentee to account may become necessary if you know that your mentee is capable of a better performance or if they haven't honoured an agreement. Challenge in mentoring, as in all other social interactions, can be uncomfortable. At the outset it is essential to understand that ***effective challenge requires that you have first established trust.*** What follows is adapted from the account by Blakey and Day (2012) to illustrate the conditions for different mentoring situations. Progress on the mentee's performance depends on both the strength of the mentor's challenge and the level of trust that the mentee places in the mentor. Consider a six-cell matrix with a scale of low to high trust

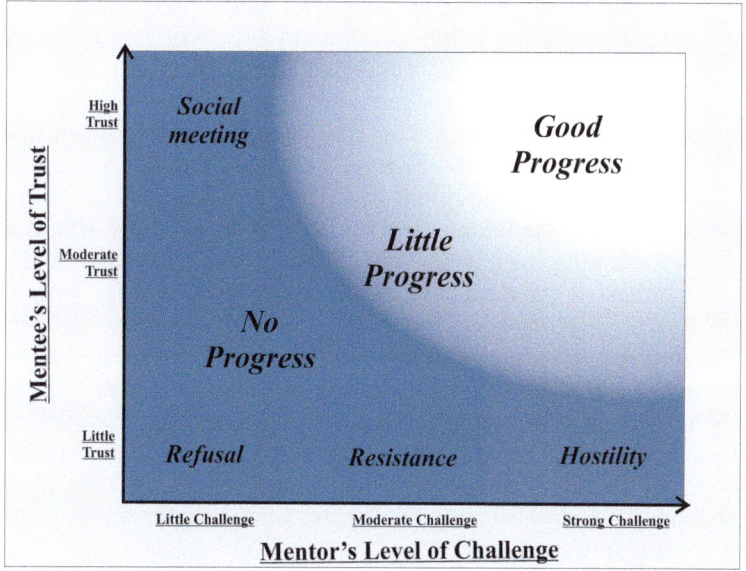

Figure 3.2 The trust – challenge matrix.

on one axis and weak to strong challenge on the other as shown in Figure 3.2.

1   *No progress* is really possible when the mentee does not trust the mentor. That is natural at the very beginning of a mentoring relationship and is why the mentor must work to gain the mentee's trust from the outset. Nor will performance be improved when a mentor has provided little or no support or even has acted to create distrust. But even with moderate and high trust, no progress can be made against resistance or underperformance if the mentor does not challenge the mentee. The pair will develop a socially comfortable but static mentoring relationship.

2   **A little progress** can be made under conditions of high to moderate trust with moderate challenge. Higher levels of trust require are created as mentoring progresses. This means that even dysfunctional mentoring can work to a limited degree provided that the mentee trusts the mentor regardless of his/her ineptitude. It probably works because the mentor's congruence and willingness to help is evident and the mentee's intrinsic self-motivation takes over.

3   **Good progress** in cases of resistance and underperformance requires high trust by the mentee and strong challenge by the mentor. The mentor's actions in this case have been picturesquely called 'the loving boot' (Blakey & Day, 2012). The mentee's high level of trust in the mentor's good will and professional skill means that the strong challenge does not damage the relationship or reduce the level trust in future. In fact, strong challenge under conditions of already high trust, if successful, can act to increase trust even further. At an extreme, this case risks developing a dependency in the mentee, and once again emphasises the mentee's need for self-responsibility.

The first step in creating accountability is actually taken right at the beginning of mentoring, long before any challenge is required. During the process of contracting at the start of a mentoring engagement you should ask if your mentee wishes you to hold him/or to account in the future of whatever is the issue being approached (see Section 4.2). Case History 3.3 illustrates what happens when advice is given before trust is developed.

CASE HISTORY 3.3 – THE CURSE OF A POORLY THOUGHT-OUT E-MAIL (JWA)

A mid-career manager was introduced to me in the side-lines of a scientific meeting. We got on well and had an amiable conversation. Not long afterwards she contacted me by email asking for feedback on a draft document – a plan for training scientists in her organisation. The plan was in the earliest stages of development and had not been fully thought through. Some haste was required because she had upcoming meetings on the subject and so, without thinking too much, I replied with the honest feedback that I thought had been asked for. I simply pointed out difficulties and suggested alternatives. Although I was sincerely trying to be helpful, in her response my colleague clearly showed that she was offended by my comments. In hindsight I should have declined to comment on the draft document until we had established a stronger and more trusting professional relationship that included a great deal of more positive interaction. Since our e-mail exchange it has been hard to recover our initial amiability and I regret that we have lost touch.

EXERCISE 3.2 – ESTABLISHING RAPPORT IN A MENTORING SESSION

Read the following before your next mentoring session or even a business meeting with a colleague. Plan, hold the meeting and try out the recommendations.

Preparation – Centring

This is the most important step. Our conscious intentions unconsciously guide our behaviour. Before the session begins, take five minutes by yourself to prepare. Sit quietly, ideally in a place where you won't be interrupted. Sit upright and relax, noticing your own body, consciously easing any muscle tensions you sense. Notice your breathing, its depth and rhythm, but do not try to alter it. Just watch. When you're feeling calm bring thoughts about your mentee to your conscious awareness. Recall previous sessions. Ask yourself, "What would be the best outcome for this meeting? Do I really want to help this person? In what ways are we similar? What interests do we share? What do I admire and/or respect about him/her?" If you sense any resistance in yourself, reframe your intentions to find some useful mutual purpose; something that is both honest for you and likely to be helpful for your mentee. If you do not do this, your mentee may sense your incongruence in the session.

Begin the session – Notice what is going on within yourself

Listen to the mentee at the same time as paying attention to the substance and quality of your self-talk and to your own physical sensations and feelings. Then reduce any tension and frustration by breathing gently, consciously allowing your body to relax.

During the session – Notice the degree of matching between yourself and your mentee

If you are matching, assume rapport has been established. Just keep going with the session.

If you are mismatching, say because of simple distraction, then re-focus on the intentions you framed during preparation and centring. You usually won't have to do anything explicit. Just get back on track.

If your mentee is behaving incongruently or obviously mismatching you don't try to ignore it. It is a sign that he/she is internally distracted by a thought that is taking him/her away for your shared dialogue. You will need to attend to that before you can achieve any useful mentoring. Note the important difference between observation and interpretation. You cannot assume you know what is happening to your mentee. It is often helpful to the mentee to make your observations explicit.

After the session – **reflect on the outcome and how it was achieved.**

What were the feelings, actions and non-verbal component of the interaction?

How well do your feelings and impressions match the outcome?

Did you notice anything that you were not aware of previously about the way you carry out meeting like this?

3.5 SELF-MANAGEMENT

O wad some Power the giftie gie us
To see oursels as ithers see us!
It wad frae mony a blunder free us,
An' foolish notion:
What airs in dress and gait wad lea'e us
An' evn devotion!

"To a louse", Robert Burns, 1786

3.5.1 Interpersonal communication skill and self-awareness

Creating awareness is the primary objective for mentoring in professional development. Mentees cannot begin to learn about any subject until they are aware not only that the subject in question exists but also that they understand their own level of ignorance about the subject. Thereafter it is reasonable to hope that intrinsic motivation takes charge and they become fully aware of the subject, eventually arriving at an expert level. Either that or else they make a deliberate and rational decision not to bother and to accept the consequences.

The ability to create awareness relies on the mentor having a relatively high level of interpersonal communication skill and self-awareness himself or herself. Within groups of STEM professionals a sophisticated level of interpersonal skill is highly valued, not only between peers but also by stakeholders of all kinds – clients, administrators, fund holders and the general public. How a mentee learns to become a fully-fledged professional is strongly influenced by the mentor's own interpersonal communication skill. Therefore, it is hard, perhaps even impossible to mentor effectively unless we have a clear and conscious awareness of our own personal communication style.

Self-awareness is therefore an essential attribute of a good mentor. The ability to help another person to create their own awareness involves taking multiple sources of information and presenting them in a way that helps that person to gain understanding of the connections. It expresses insights in meaningful ways, addresses underlying concerns and encourages the learner to enquire on their own account and so increase their awareness and clarity to create new thoughts, beliefs and perceptions. Reflection on increasing awareness of the outside world ideally leads to increasing awareness of one's own interior world, of our own values and beliefs, strengths and weaknesses. The case history given at the end of Chapter 1, that of the recent graduate who learns to think for himself, is an example of the awakening of self-awareness in a mentee.

Among the most helpful general works on skilled interpersonal communication and self-management is an article by the influential management thinker, Peter Drucker in which he says, "Success in the knowledge economy comes to those who know themselves; their strengths, their values and how they best perform" (Drucker, 2005). Stephen Covey's 'Seven habits of highly effective people' became famous not only for its simple wisdom but also because it is more thought-provoking than the average self-help book (Covey, 1989). For a more rigorous academic approach, including extensive references in psychology and social science, Hargie and Dickson (2004) and Guirdham (2002) allow a closer look at this complex subject.

3.5.2 Emotional intelligence

In what became known as the Age of Reason, the 17th and 18th centuries in Western Europe saw the first flowering of science, engineering and technology. Our professions are essentially founded on rational thought. By long-standing convention, scientists and engineers are discouraged from using personal pronouns and prefer the passive voice as if there were no human agents in our work. Emotion it seems, has no place here. Although all substantive argument within STEM fields should be entirely rational, it is obvious that we, as the all-too-human practitioners, do not always behave reasonably. Anyone who has attended a scientific meeting in which opposing opinions were expressed, often on trivial or arcane matters, will have experienced just how much emotion can colour the arguments. In fact, the majority of scientists and engineers do not work at a job so much as follow a vocation with a passion that dominates their personal identities. Emotion, far from not being present, actually covertly dominates all our work.

Emotional intelligence, an expression coined by Daniel Goleman, was brought into popular currency in the 1990s. It refers to the capacity for recognising our own feelings and those of others, for motivating ourselves and for managing emotions well in ourselves and in our relationships (Goleman, 1996). Goleman's insight is that our level of emotional intelligence does not stay fixed through our lives. Rather, emotional intelligence is a learned cognitive skill like empathy (which it includes), one that can be developed as we progress.

In what is closely related to emotional intelligence and self-awareness generally, 'Understanding Self' is the first of eight competence categories in mentoring, according to the European Mentoring and Coaching Council (EMCC, 2020). By this is meant awareness of one's own values and beliefs and of how these affect one's mentoring practice. It is closely similar to, 'Coaching Presence', as described by the ICF (2020) as their fifth competence category. That is described as being fully conscious and present in mentoring, being focussed, open, grounded and confident with the ability to manage one's emotions. Whatever the exact definitions, self-awareness and emotional intelligence are key skills and fundamental attributes of an effective mentor.

3.5.3 Non-verbal communication

A great deal of the feelings and attitudes we communicate are carried by NVC. That is the general term for body language, paralinguistics (see below), symbols in dress and context and all other forms of personal expression not carried by spoken language. NVC functions as a compliment to verbal communication. For example, it regulates conversation in allowing us to take turns and it conveys emotions and such social factors as dominance and identity. NVC is mainly, but not wholly, unconscious. As the previous discussion on establishing trust showed, when what we say contradicts how we feel, then our incongruence is revealed by our NVC and is referred to as 'leakage' (see Section 3.4). The subject is covered in some detail within Hargie & Dickson (2004)

As a mentor, you need to be consciously aware of your own characteristic NVC as well as that of your mentees. With NVC both you and your mentee are communicating your mood and attitudes, such as approval and disapproval, to each other. 'Pacing' is the subtle art of influencing by first consciously matching another's body language and paralanguage in order to gain rapport. Pacing is then followed by 'leading' in which new ideas are put forward with the intention that the person who has been 'matched' can be more easily influenced. Obviously, pacing and leading could be practiced or experienced as a form of unscrupulous manipulation.

Paralinguistic communication includes aspects of vocalisation in addition to the meanings of the words spoken. Experts say that the basic paralinguistic modalities are volume, articulation, pitch, emphasis, and rate (Hargie & Dickson, 2004). If you pay close attention to your mentee's paralinguistic communication, you can get a clearer sense of what they feel and what their attitudes are to the subject in hand. For the most part, this is a natural, normal and mainly unconscious skill that we all possess, but conscious awareness helps to increase our understanding. As only one of many paralinguistic modalities, consider for example, emphasis. By stressing specific words or phrases in speech we can create distinctly different meanings that may not be apparent in written communication. For example, in response to a suggestion, if your mentee says, "I can't do that here", note how unconsciously placing the stress on different words carries unspoken meanings:

- "<u>I</u> can't do that here", implies personal incapacity that does not apply to other people
- "I <u>can't</u> do that here", implies shock at the very idea, definitely not
- "I can't <u>do</u> that here", implies I might be able think it or find some other way
- "I can't do <u>that</u> here", implies shock or maybe other actions might be possible
- "I can't do that <u>here</u>", implies doing it somewhere else is possible.

Other forms of NVC that mentors need to be aware of are those mediated by choices in physical appearance, clothing, accessories, context and environment. Imagine how ineffective a mentoring session would be when the seating is arranged so that the mentor sits behind a large desk in a comfortable chair while the mentee sits opposite on a lower and less comfortable chair. The mentor wears an expensive suit while the mentee is dressed in casual old clothes. During the session the mentor breaks off frequently to attend to interruptions, phone calls and visitors. In this situation everything communicates the mentor's attitude that he intends to dominate and that the mentee's issues are of lesser concern than his own. A poor outcome for the mentee is guaranteed.

3.5.4 Using your intuition

Most scientists and engineers would not admit to using intuition. It has a flaky, irrational connotation with echoes of superstition. The dictionary defines intuition as, "Immediate apprehension by the mind without reasoning" (OED, 1964). So, looked at another way, we unconsciously use intuition in many quite normal circumstances. In social settings, for example, we all unconsciously observe and react to other people's NVC. Long experience in professional practice gives us the ability to arrive at accurate conclusions in our own specialisms without going through conscious and maybe long-winded rational cognitive processing. Intuition is more usefully thought of as a kind of heuristic, a rule-of-thumb or acceptable short-cut born out of extensive experience.

In mentoring, it really helps to use your own intuition about what the mentee is thinking *but* do so only with caution. Be aware that you cannot assume that what your intuition seems to be telling you is correct in reality. Therefore, always check your intuitive conclusions with your mentee. If you're wrong, then your mentee has the opportunity to correct you and so you share information about what is really going on.

e.g. "I see you shifting in your chair and looking away. I feel that you might be unhappy about the outcome of our discussion?"

3.5.5 Non-directive mentoring

Self-awareness gives us the ability to be flexible in our approach to our mentees without being in any way incongruent or manipulative. We can vary our styles of interacting, for example, as between formality *vs.* informality, simplicity *vs.* elaborateness, task-orientation *vs.* people-orientation and directness *vs.* indirectness.

Arguably the most important flexibility is to decide whether to tell your mentee what to do or to help him/her think out a solution. This decision is what in counselling is called adopting a directive or a non-directive style. The difference between directive and non-directive styles is comparable to the difference between a parent–child conversation and a conversation between two adults. On the one hand, the parent gives directions because the child does not have prior experience and thinking skill sufficiently developed to solve the problem alone. On the other hand, an adult conversation respects the experience and autonomy of the other adult because they already have some prior experience and thinking skill. The mentee's own responsibility to learn is greater than the mentor's to inform. By declining to give direction you are encouraging a rational, unemotional discussion between equals. The coaching and mentoring institutions require a non-directive style to be used wherever possible. Starr (2014) gives a detailed account of directive and non-directive language.

Both styles can be appropriate and useful in different circumstances. There is nothing intrinsically wrong with being directive. For example, it is normal for managers and consultants to be directive. They are trained to take the lead, determine the form, pace and content of a meeting, call on others for responses and dominate the interaction by delivering the majority of the information and opinion. Although the requirement to help and advise might appear the same in both mentoring and management, mentoring has an individual rather than organisational purpose and uses a different interpersonal skill set and approach. Mentoring is definitely not management or consultancy.

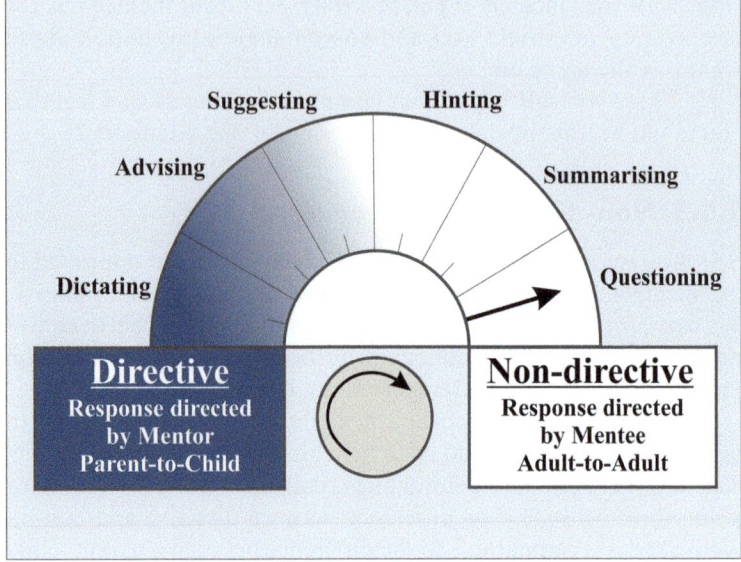

Figure 3.3 A spectrum of directive to non-directive mentoring responses.

In practice the distinction between directive and non-directive styles is not binary. A mentor's choice of how to respond to a request for help or advice lies somewhere on spectrum between two extremes, as illustrated in Figure 3.3.

Say, for example, a mentee comes to you and asks, "What should I do with this set of results?" Your options for responding range from direct to indirect and might be as follows:

- Dictating – "Load the results into the application and I'll show you what to do"
- Advising – "Taylor is the acknowledged authority on this subject. Read his 2012 paper"
- Suggesting – "In my opinion Taylor's method is the best, but you can find other options"
- Hinting – "It looks as if you haven't read the literature on this?" [effectively begs a question]
- Summarising – "Here's what I heard you say about these results …" [listen & discuss]
- Questioning – "What do you think you should do?" [questioning sequence follows].

CASE HISTORY 3.4 – DEALING WITH ANGER (IAN M. GRAHAM)

The medical curriculum has traditionally been based on didactic lectures and an apprenticeship approach in which the student attends ward rounds, clinics and investigations. The degree of mentoring is informal and unstructured. The slow realisation that 'talking head' lectures are not very effective has led to greater interactivity, and a shift from rote learning to clinical problem solving. Dealing with bereavement is now taught in most curricula, but not much attention is paid to mentoring with regard to either communications or dealing with anger.

Joe was well known as a diligent intern, often working long beyond his duty commitments. It was 8.30 p.m. when he asked to talk with me. After 12 coronary angiograms and a pacemaker, I was almost as tired as he looked, but he was clearly distressed so we took a coffee into my office.

He said, "Prof, Admin gave me this letter and asked me to respond but I don't know how to".

I thought that this admission was quite a good start. "OK, first tell me what happened, then let's look at the letter".

Joe told me he had been working 20 hours straight and had more stuff to do. The patient had had a mini-stroke after an angiogram which scared him badly – he lost his speech and the power in an arm but later recovered 100%. The family cornered Joe in the corridor, said it was impossible to get to see anyone and demanded to know 'What went wrong'. They said we 'Must be covering things up' and that they were determined to 'Get to the bottom of it and find out who was responsible'. Joe tried to explain that the patient had had a micro-embolus, how it could cause a hemiparesis and aphasia and was about to explain more when one of them shouted at me 'Oh, for God's sake you're trying to tie us up with medical gobbledygook. We are going to take this further'. Joe admitted, 'To be honest, I lost it. I had a million other things to do and I probably shouted something like, "Fine, do that, now get out of my way and let me get some work done"'.

Again, I felt that his honesty was a hopeful sign. I said "OK, maybe not so smart but I remember being wrecked and feeling harassed too. Let's look at the letter".

The family demanded a full and comprehensive explanation and were prepared to, 'Take further action'. But frankly, compared to some letters I've seen, it could have been worse. Although the situation was serious and the family were clearly still very angry, its tone and wording suggested that they might be open to discussion. Key issues were-

- The staff seemed rushed off their feet and it was almost impossible to get to talk to anyone.
- Their father, the patient, was terrified at what happened and so were they. He couldn't speak for some hours, his face was crooked and he was dribbling, and his arm wasn't working.
- When they finally got to speak to the intern, they couldn't understand anything he said. They had no idea if there was permanent damage or if the problem might recur.
- The intern then shouted at them.

We paused and thought a bit. I said that I had 'been there, done that' several times early on in my career. I also said that he had helped by being open and not defensive but also that the letter seemed to me to be equally honest. In the discussion that followed we explored the following issues and in fairness he contributed as much or more than I did:

- The system is unfair, with ridiculously long hours, and we are expected to behave like angels and not to have normal emotions.
- We are not trained to deal with anger.
- Complaints may arise because something bad was done, because the relative is on a mission and nothing will satisfy them, or simply because of bad communications by us. I suggested that, in my experience, the last was by far the most frequent problem.
- We are not taught communication skills. From the extensive literature on this subject, core issues are the use of medical jargon,

information overload, our underestimation of people's health literacy and the simple fact that frightened people cannot easily hear or comprehend.

So what next? We considered a letter, but eventually concluded that it would be best to meet the family as soon as possible, promise them ample time, hear all of their concerns, and go through the case notes with them step by step. I asked him if he would like to do this, or prefer that I handled it, or we could do it together. He opted for the latter. To his great credit, and as a testament to his growing maturity, he took responsibility for simply telling the family that he was exhausted and that he had not explained things properly. At this kind of meeting, the initial body language can be very tense, sometimes (and reasonably) accusatory. The intern's palpable sincerity, almost tearful once or twice, contributed in no small way to the draining away of tension. I did not have to spell it out that it was a learning experience for all. The family thanked us, but especially Joe.

3.6 GIVING CONSTRUCTIVE ADVICE AND STRUCTURED FEEDBACK

As a mentor you'll often be called on to give advice. However, the word 'advice' can mean different things in different contexts. For example, in the situation where an employee is up for reprimand and the boss says to him, "Let me give you some advice", the word 'advice' is used in the sense of giving performance feedback. The boss's verbal tone and the context may even indicate a certain level of threat so that you have no real choice about how to respond to the 'advice'. Contrast that with the situation where a patient who is worried about some symptoms seeks advice by attending a doctor for a consultation. The patient in this case is asking for a professional opinion. In giving an opinion the doctor is not in a position to insist that the patient takes the advice on treatment. However, strongly the doctor feels, he/she can only recommend treatment. Similarly, there are two contrasting situations in which you, as an expert in your field, might be called on to give advice to a mentee.

One, similar to that of the doctor, is when you are called on to give constructive advice based on your professional opinion. The other situation is when you have to give your mentee advice on performance. This should be in the form of structured feedback. The two different forms of advice are discussed separately below.

3.6.1 Giving constructive advice

The main difference between mentoring and professional executive coaching is whether or not you give advice in the sense of a professional opinion. Executive coaches are trained *not* to give advice. An executive coach's primary skill lies in facilitating their client's own thinking and in helping them to create their own solutions. In any case, coaches are not usually expert in the same field as their clients and so not in position to advise. In contrast, a mentor is, by definition, an expert in the same field as the mentee and so is qualified to give a professional opinion. As scientists and engineers in situations other than mentoring we are trained to give technical advice. So, when a mentee comes to us looking for advice, our default response is to give it. However, giving direct advice is usually not the best response in mentoring. Among the commonest reasons mentees might ask for advice are a simple desire to save mental energy, a lack of confidence or an unwillingness to shoulder the risk entailed in making a decision. All of these reasons reveal an underlying disinclination to accept professional responsibility. So, instead of giving advice to mentees a much more effective response is to set tasks, such as reading a textbook or paper that will enable them to get the information for themselves. You can then support them by arranging a subsequent meeting and helping them think though the problem with the benefit of the new information.

However, occasions do arise when responding to a request by giving direct advice could be necessary or effective in mentoring. Such occasions might be when mentee cannot acquire the knowledge within any reasonable timescale because he/she is out on site, in the field or in a clinic without access to libraries. It can also happen that a little constructive advice is necessary in the middle of a long technical discussion in order for the mentee to make progress. Another reason might be that the knowledge required is so specialised or advanced such that they could not work out the problem unaided. That would be the situational equivalent of attending a doctor when the patient has no medical knowledge. *It should therefore be a relatively rare event within the same profession.*

Giving constructive advice means exercising sound judgement, often mistakenly thought of as a quality we either have or don't have. Exercising sound judgement is, actually a professional competency in its own right, a learned skill. The greater part of science, technology and engineering training involves the study of methods for collecting evidence and how to apply general theories, practices and rules to the evidence in order to arrive at conclusions and opinions. All competent professional judgements or opinions should be explicitly based on evidence. The best role models for how to give advice are to be found among the legion of experienced and well-regarded family doctors and solicitors. Garvin and Margolis (2015) give a very helpful discussion on how to give professional advice generally. The following recommendations, and the flowchart in Figure 3.4, are adapted for the special circumstances of giving advice in mentoring.

Gratuitous advice is advice that has not been asked for. Even if it has been given with the best intentions, it will often be ignored and may even cause offence. In a mentoring session, if you realise that your mentee needs some piece of information in order to make progress, then *do not offer the information or advice without first checking the mentee wants to hear it.* Even if the mentee agrees that they do actually want to receive the piece of information, a more effective approach is usually to engage in a questioning sequence with discussion that results in the mentee realising that they do actually need that specific bit of information.

If you have decided to give advice and feel that it may take some time, say to support a decision about career transition, then first make preparations. Arrange a private space where you will not be interrupted and limit the time to say, 30 minutes. Make sure that you are the best person to give the advice and that you have the required knowledge and expertise. Also explain to the mentee that that your role is only to offer guidance while his/her role is to make the actual decision. Begin with a questioning sequence to achieve a mutual understanding of exactly what advice is needed and all the material details.

The next step is to offer as many reasonable options as you can. Give your reasoning and personal experience and ask open questions to help the mentee assess each option. Then, without haste or pressure, encourage them to close in on a decision. Make sure no

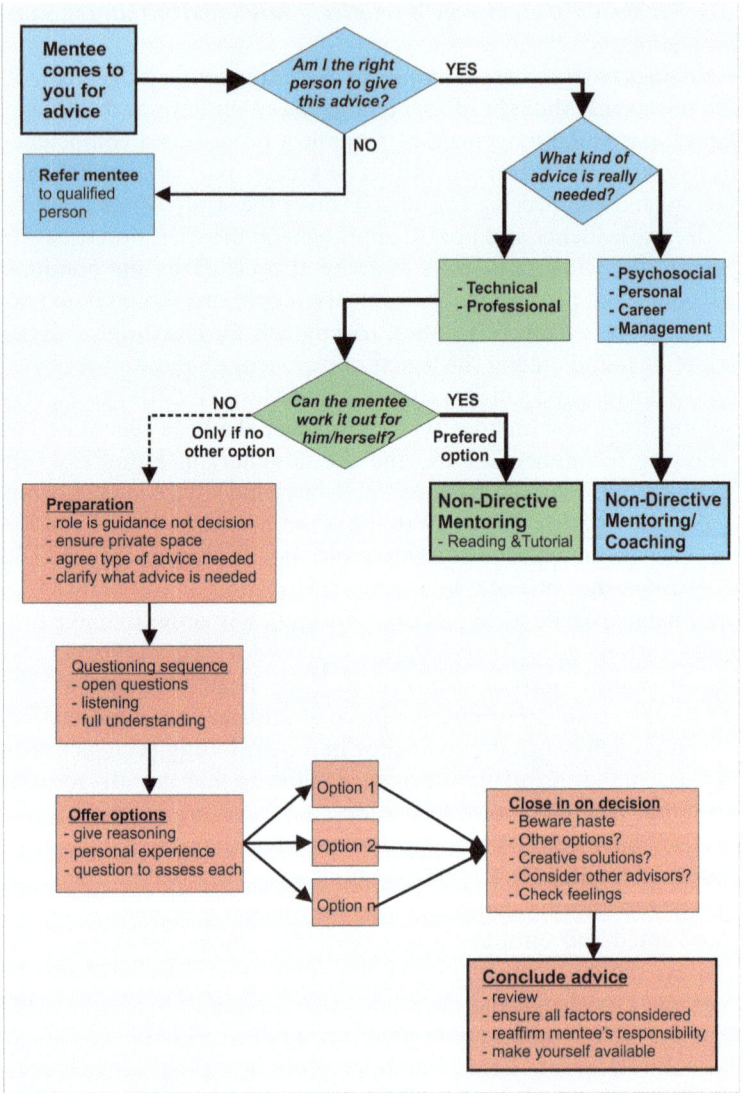

Figure 3.4 A flowchart for giving advice

options have been missed out or assumptions unchecked. Reaffirm your mutual understanding that the decision is solely the mentee's. A flowchart summarising these steps is given in Figure 3.4.

EXERCISE 3.3 – GOOD AND BAD ADVICE

Work with a partner and take turns. Focus on your intentions to:

a Understand the process of giving effective advice, and to,
b Practice the basic mentoring skills – listening, questioning, building trust and flexible approach

Describe an experience of giving or receiving advice while your partner listens and asks occasional clarifying questions. The actual experience is not important. Your objective is to understand the process.

Take 15 minutes each and switch roles, then take a few minutes to discuss what you learned:

- *What was the context and background?*
- *Who was involved (including third parties - false names if needed)?*
- *What actually happened during and afterwards as a result? Deconstruct the exact process.*
- *What was your reaction? How did you feel?*
- *What made the advice effective or ineffective?*
- *How could the advice and its delivery have been improved?*

3.6.2 Structured feedback

In this section, the word 'advice' is used in the alternative sense of *feedback on personal performance or behaviour.* This kind of feedback can be either positive, as in as expressions of approval and congratulation for something well done, or negative, as in expressions of disapproval and criticism for a mistake or poor performance. Positive feedback is essential in giving encouragement and in reinforcing useful developmental behaviour in the future. Negative feedback, if it is properly given, can be constructive in making us think more clearly and in promoting a positive change. A positive behaviour change requires two essential conditions:

1   The receiver must trust the opinion of the person who gives the feedback, and,
2   The receiver must feel that the feedback is appropriately delivered and constructive.

If these conditions are not met and the receiver perceives the feedback as an insult or a threat, then he/she can experience strong negative emotions and confusion, maybe so overwhelming as to block action. In contrast, well-delivered feedback can enhance rapport, create trust and so develop an open and honest relationship with the person who gives the feedback. Skilled delivery of negative feedback can lead to the receiver making a significant career step. He/she may even feel gratitude.

Ten guidelines for giving constructive feedback are:

1   *Give feedback in person*, face-to-face, making eye contact. Negative feedback requires privacy and freedom from interruption. Positive feedback is strengthened if others are present.
2   *Prepare* for how you want to be heard, make sure of the facts and organise your message.
3   *Be honest with yourself about your motives* before giving feedback. Ulterior motives (*e.g.* revenge, dominance and manipulation) will inevitably come through in the feedback and damage the relationship.
4   *Hold the intention to work for a constructive and positive outcome.* Intentions created consciously later guide our actions unconsciously. Framing a positive mind-set is more likely to produce a positive outcome.
5   *Be objective.* Distinguish clearly between the factual evidence and the conclusions you draw, between observations of behaviour and interpretations of what you think it means.
6   *Recognise that you could have reached the wrong conclusion.* Give the opportunity for reply and allow for correction in the light of new information.
7   *Maintain a respectful manner.* Even if you don't agree, the receiver is entitled to his/her opinion. Manage your own NVC carefully and watch the recipient's responses.
8   *Never personalise.* Focus on the facts and observations of the behaviour rather than on the person. Name-calling and personal criticism will only increase tension and anger.
9   *Don't generalise.* Make feedback specific and detailed. – *e.g.* Instead of saying, "That was good a report". It would be

more effective to say, something like, "Your report gave a really clear explanation and that allowed me to understand this problem fully".

10 *Make your comments clear, concise and relevant.* Address a difficult subject directly, being firm and polite. Avoid waffle and lengthy justifications. Don't mix good and bad (the so-called "sandwich") in a way which might confuse.

Most authoritative coaching textbooks, including those of Starr (2014) and Hawkins and Smith (2006), give detailed accounts of giving various kinds of feedback. In approaching the subject from the viewpoint of holding someone to account Patterson *et al.* (2013) offer a great deal of useful information on this tricky subject.

CASE HISTORY 3.5 – YOU NEVER SAID THAT BEFORE (IAN GRAHAM)

As Head of Cardiology, I was rather pleased with our multi-disciplinary monthly team meetings. Everyone was encouraged to say their piece – even if the seniors perhaps talked too much and listened too little. We were all over-stretched trying to cope with excessive demands with too few staff and insufficient equipment. This month had been particularly tough. Many people were exhausted. We went through all the issues and at some stage I thought I would boost morale by saying, "But, guys, everyone knows you are fantastic, always make the extra effort, always go the extra mile". The most junior technician said quietly, "Sir, you never said that before". I felt about one millimetre tall. Had I really been so crass as to assume that they would know what I never actually said? I hope I learned that credit has to be expressed.

EXERCISE 3.4 – ROLE PLAY GIVING STRUCTURED FEEDBACK

Setup: Work with a partner sitting in a quiet place where you will not be interrupted. Each think of a difficult conversation you have been avoiding or one in the past that did not go well and you'd like to think about what you could have done better. Your partner will act as the receiver of your feedback and, in turn, will give you feedback on how it went from their perspective. Allocate an agreed amount of time, say ten minutes each and then switch roles.

Note that this exercise has the potential to be emotionally upsetting. Choose a simple feedback scenario; one that does not have a very high emotional loading. Do not choose one that in real life actually involves your partner. As before, the substance of the scenario is not important. The object is to understand the method.

Plan: Take two minutes to explain the circumstances to your partner. Explain your intentions, motives and purpose. Plan the location, sequence, atmosphere and outcomes.

Begin the role play: Without chat or pleasantries, start directly by saying what you want to talk about and why you are addressing the subject. It may be that that a colleague has asked for feedback or that your role as a manager makes you responsible to give feedback. You may ask permission to give negative performance feedback, e.g. "We have already agreed that you have given me, as your mentor, the responsibility of holding you to account. May we discuss your reasons why you have not submitted your Chartership application?"

Say what you intend: Say why you are giving this feedback and the outcome you hope for, e.g. "My intention here is to help both of us to learn from this experience and to see your performance improved". In order to defuse anxiety and tension, it often helps to say what you do *not* intend, e.g. "There is no question here of bringing a disciplinary procedure". Maintain a respectful and serious manner and tone. Make sure your behaviour is congruent with your intentions.

Provide the evidence: Provide the objective evidence or observation of specific behaviours or actions. It is important not to minimise but neither should you sound angry or judgemental. Remain objective and detached, e.g. "I heard you say ...", "I saw you ...", "On page 15 of your report you say..."

Give your interpretation: Then say how you have interpreted that evidence or what conclusions you drew; "...and so I thought ...", "That made me feel ...", "So I expected ..." You must own your conclusions and not generalise or adopt a victim position: Do not say, *e.g.* "Everybody says …", "That was awful…" or "You made me feel …"

Invite a response: Next give the opportunity for response and correction. Remember there could be reasonable mitigation that you didn't know about; *e.g.* "Have I understood correctly?" or, "How do you see it?", or, "What was your intention?", or, "What will you do now …?"

With negative feedback, you can encounter strong resistance at this point which could become acrimonious. Maintain your personal intention to stay objective, calm and respectful.

If necessary, restate your intention and say what you do not intend. Bring the focus back to your intended outcome. Also, look out for, and respectfully challenge any discounting (minimising) of what is actually your carefully considered opinion. With negative feedback, a recipient may say, *e.g.* "Oh come on! It was only a trivial matter". Alternatively, when praised, a recipient may say, *e.g.* "It was nothing really". In both cases, your considered opinion has been discounted.

Clearly give the consequences: Finally, say what will happen next. Then close, *e.g.* "The consequences are…", "This must mean that …", "We'd therefore need you to …" If you are asking for a specific behaviour change or action make clear exactly what that is and, importantly, check that they have the knowledge and the means to accomplish it (*i.e.* the self-efficacy); *e.g.* "Have you done this kind of work before?", "Do you know how to carry out Read's test?" If feedback is positive then note the satisfactory outcome or reward, *e.g.* "Your work over this last week has been a major advance for our project. The CEO has expressed his satisfaction". Express gratitude, admiration or acknowledgement as appropriate. If some action is required of you, make sure you follow through.

3.7 A BROADER VIEW OF MENTORING SKILLS

The great majority of executive coaching skills are also used in mentoring. There are many more skills than the five we have just described. They are all to be found described within texts listed in References. Each of the major coaching and mentoring institutions lists and describes coaching skills as the competencies required for professional accreditation. The Association for Coaching lists nine competencies (AC, 2020). The ICF describes 11 competencies distributed between four groups: (A) setting the foundation, (B) co-creating the relationship, (C) communicating effectively and (D) facilitating learning and results (ICF, 2020). Similarly, the EMCC lists eight categories of competencies. They all say much the same thing but in different ways. The EMCC competency categories may be taken as representative:

1   Understanding self
2   Commitment to self-development
3   Managing the contract
4   Building the relationship
5   Enabling insight and learning
6   Outcome and action orientation
7   Use of models and techniques
8   Evaluation

Competency groups (1), (2) and (8) refer to the coach/mentor's own commitment to the process, including self-management and also the application of principles which are discussed in the next chapter. Competency group (3) Managing the contract, is central to the whole process, as described in the next chapter. Competency (4) Building the relationship, includes the ability to build trust and also language skill appropriate to the person and the situation. Group (5) Enabling insight and learning, involves skilful questioning, active listening and feedback, as well as offering one's own perspective in a way that gives the mentee choice. Group (6) Outcome and action orientation, involve clarification, planning and goal setting and maintaining focus on solutions. Finally, group (7) Models and techniques are discussed as part of the mentoring process in the next chapter.

CASE HISTORY 3.6 – WHAT I HAVE LEARNT AS AN ACADEMIC SUPERVISOR (GUS HANCOCK)

My graduate supervisor and I were both academic virgins as far as the mentoring process went. I was his first graduate student, and my undergraduate experience had consisted of straightforward instruction. But I learnt a great deal on how to be a mentor because mine was excellent at it. Some of the characteristics that I experienced I hope will pass on.

1 *Positive reinforcement* – My graduate supervisor was really pleased for me when I devised and carried out a new approach to solving a previously enigmatic problem. He had guided me there and given me the tools (and confidence) to go further. Many of my graduate students have shown similar independence of thought.

2 *Time management* – Academic supervisors generally appear to be busy or harassed, but such an appearance is no excuse for putting off the mentoring process by unnecessarily re-scheduling meetings or failing to read submitted work within a reasonable timeframe. Don't take on the mentoring task if you are going to procrastinate.

3 *Reward* – Part of my job was to get the research group members to an enjoyable conference once a year. And these are experiences that the group remember vividly.

4 *Feedback* – Part of all graduate students' training nowadays is the ability to give a decent presentation, and these featured heavily throughout my career as a mentor. Give positive feedback straight away both from the mentor and from others within the group listening to the presentation.

5 *Honesty* – I have had the privilege of mentoring many graduate students who have gone on to successful academic careers. Others have wanted to but needed a bit of advice about their prospects when I thought them to be thin. Don't flannel.

And finally, *some thoughts on the mistakes* that I have seen (and made) both as a supervisor, a tutor for graduates in an Oxford College and as Head of Section in the University. When I started an academic career in Oxford 45 years ago there was little training in how to teach. Now the emphasis has changed and professional training is widely available (and more recently mandatory for first time supervisors). Some mistakes I have seen:

1 *Poor communication* – I have had to intervene between supervisor and supervisee because of little feedback, goals not properly set, a lack of understanding of what the supervisor wanted and what the supervisee could achieve. Generally the intervention came at a late stage, with much damage already done and mutual trust evaporating.

2 *Remoteness of the supervisor* – Common in very large research groups, where supervision is delegated to senior postdocs and tends to be instruction rather than mentoring. I have come across several very clever young people who have been disappointed with the fact that they did not get the mentor they were expecting, and for some it led to a move out of research.

3 *Too much availability* – From the remarks above this one seems far from being a problem. I made this mistake early on, by stating that I could be available for advice at any time. I was overwhelmed by one graduate student asking for reassurance about trivial matters which could have been solved with a bit of thought beforehand. I learnt to recognise the knock at the door and shamefully on one occasion hid behind my desk as the apparently needy mentee came to look in. I eventually explained that I was going to introduce some ground rules after that, particularly to get the student to think more clearly about the perceived problem instead of simply asking me for the solution. It worked better that I could have hoped, and the result was a very successful doctorate.

Chapter 4

Mentoring principles and process

4.1 MENTORING PRINCIPLES

4.1.1 Principle 1 – The primary role of mentoring is to create awareness

The primary role of a mentor is to create awareness of the possible choices and help the mentee take responsibility and ownership for the decisions they make, in this way to find authorship in their role, relationships, tasks, activities, in fact their whole career (Leary-Joyce, 2014, with permission from the author).

The statement above is a slightly amended version of the same statement about coaching principles written by John Leary-Joyce, one of the best-known exponents of the Gestalt School of coaching (Leary-Joyce, 2014). The only amendment here is in substituting the words *"mentor"* for *"coach"* and *"career"* for *"life"*. The German word 'Gestalt' is a descriptive noun meaning something that is complete, a whole or forming a pattern. Gestalt psychologists and coaches employ a process of helping clients to become aware of the subject under discussion in such a way that they gain a holistic understanding of their choices and so are enabled to determine their future actions in the broader context of their whole lives. When a mentee comes to you, it is because some aspect of their professional practice has become the focus of their attention. Your job is to help them to become fully aware of the whole range of connections to that aspect of their practice in such a way that they want to use their newly acquired knowledge and skills to develop their career.

Creating awareness, or the Awareness Principle, carries with it two significant corollaries. One is that ***effective mentoring is a skilled activity in its own right.*** There is more to mentoring than just telling people things they should know about the job. Mentoring is an interpersonal communication skill that raises awareness and motivates mentees to learn and improve. As with any other communication skill, it can be learned and developed through practice.

The second corollary is that *effective mentoring is outcome orientated*. When facing any problem, we can take one of two contrasting positions. One is a problem orientation that engages us in a circular discussion that churns through one problem after another and never finds a solution. We usually add to the misery by looking for someone to blame. An outcome orientation, in contrast, focusses on finding solutions. In the context of mentoring, an outcome orientation relates the solution for the mentee's current problem to his/her immediate purposes and future career direction. The Awareness Principle reminds us that the intended outcome of any mentoring engagement is for the mentee to gain knowledge and skill to be used in practice. Encourage your mentee to understand that he/she has the power to decide what it is that they want in any given situation. *A mentoring session is not just an enjoyable intellectual discussion; it should always involve a 'can do' attitude and conclude with a result.*

4.1.2 Principle 2 – Mentoring is client-centred

In a term borrowed from counselling, modern mentoring is 'client-centred'. In the 1950s the psychologist Carl Rogers developed a therapeutic approach which focusses on how the client sees themselves rather than how the counsellor interprets their unconscious thoughts. In effect, the method is non-directive and relies on the client's intrinsic motivation and adult understanding rather than the counsellor directing the client. Applying the client-centred approach to mentoring means that *the focus of mentoring and the primary beneficiary of mentoring must always be the mentee*. The mentor or the employer who sponsors mentoring should not impose solutions, but rather they should encourage the mentee's developing professionalism. Secondary benefits that may accrue to the employer or any others are acceptable provided there is no loss of focus on the mentee's development and no conflict of interest.

Although the Client-centred Principle may seem so obvious as to be hardly worth saying, circumstances can arise when the Principle may conflict with what others want for the mentee. Company training schemes are often set up because managers realise that their company needs to have a skill deficit rectified or upgraded in order to achieve corporate goals. In this case, the training scheme is explicitly company-centred. In most cases professionals are content to have training provided for them because it enhances their own knowledge base at no cost to themselves. Mentoring as an adjunct to formal training courses is by far the most effective way of transferring and embedding formal training within an individual's practice. In cases where mentee and mentor have been mismatched or coerced into mentoring or where one or both of the mentoring pair disagrees with the training method or its subject, then both the mentoring and the training are not client-centred and they risk being dysfunctional. To be effective any company-sponsored training and mentoring scheme must have 'buy-in' from the trainees and mentees and mentors who are required to participate.

4.1.3 Principle 3 – Self-responsibility

It is axiomatic that the essential qualities of a high-functioning professional scientist or engineer include *taking responsibility for his or her own performance and maintaining a self-motivated, independent mind-set*. Developing these qualities depends on the learner taking responsibility for his/her own professional development right from the start. It also means accepting the risks of their own independent choices.

An important corollary is that it is the mentee (*not* the mentor or the employer) who should approach a mentor to ask for help, to choose the subject of discussion and to set his/her own learning goals. At first sight, this principle might seem to be counter-intuitive. After all, the mentor is more experienced; surely he/she knows best? The clear answer must be that the mentee is an adult learner and not a school pupil. In line with the Client-centred Principle, he/she must have an adult's freedom to choose his/her own career path.

Recently graduated junior practitioners who have been spoonfed through their undergraduate training may not appreciate these conditions. They may expect that their employer simply provides them with a mentor who will train them and tell them what to do. If they are not aware of the Self-responsibility Principle, and most aren't, how then does mentoring get started at all? One answer to this conundrum is mentoring induction training for new graduates. The problem is discussed further below in Section 4.2.

Before mentoring during the scoping and contracting processed described below (Section 4.2) you should explain the Self-responsibility Principle and insist that your mentee takes full responsibility for the outcomes of whatever is the subject of the mentoring process.

Another corollary is that the mentor is *not* responsible for a mentee's failure to reach goals. The converse is also true. If you cannot be blamed for your mentee's failures, equally you cannot take credit for his/her success. The mentor is like a catalyst in a chemical reaction or a midwife attending a birth. *Mentoring is a purely facilitative activity.*

The Self-responsibility Principle also means that *mentoring must be voluntary on both sides*. Inevitably, coercion will be counter-productive. For an employer to demand that individuals engage in mentoring together is to ignore the Principle of Self-responsibility. Equally, you have no power to insist that your mentee must follow your instructions or your advice. You can only offer guidance which the mentee is free to accept or to reject.

4.1.4 Principle 4 – Intrinsic motivation

Research in experimental psychology over the last few decades has supported the belief that *human beings are naturally inclined to be creative, resourceful, self-motivated and capable of generating solutions* (Dweck, 2012). Our own inner wish to achieve whatever it is that we are striving for is more powerful than any rewards or threats from outside. The consequence for leadership is that it is more effective to support the belief that the job is worthwhile for its own sake than it is to make threats of job loss or offers of pay increases. It also follows that a lack of motivation must be caused by inhibition of this natural drive. Gallwey (2002), in an influential and widely quoted discussion on this subject, expresses the consequences of the Intrinsic Motivation Principle as a notional formula. If an individual's actual performance were to be measured, it could be calculated as his/her potential performance minus a notional measure of interference. The latter may be due to negative feelings such as a perceived inability, insufficient resources or conflicting demands. A further consequence of the Intrinsic Motivation Principle is that everyone is innately capable of achieving more than they are at present. Overcoming the interference or inhibition naturally leads to achieving more.

So, if your mentee seems to lack motivation for some task that is under discussion, then you cannot make progress by rational persuasion, promising rewards or issuing threats. Instead, you have to help them to find the cause of their inhibition, then explore it and help them find their own way of overcoming the inhibition. The best service you can provide any mentee is to help them to rediscover their own natural intrinsic motivation (see Section 5.6).

4.1.5 Principle 5 – Ethical responsibility

If mentors have the capacity to promote their mentees' professional development, then they also have a corresponding potential to cause harm. Dysfunctional mentoring can adversely affect a mentee's performance. It can foster negative attitudes, cause demotivation, stress and emotional upset and, in the worst case, damage a whole career. Although the mentee should take personal responsibility for his/her actions, the mentor never-the-less occupies a position of trust. *Mentors have a duty of care* for their mentees. The mentor must observe strict professional ethics, not only for the mentee's protection but also because the mentor is a role model. Ethical principles for mentoring described here are adapted from those required by the professional coaching and mentoring institutions.

The most basic level of the duty of care is to cause no harm. Passively maintaining professional boundaries, obeying the law and avoiding inappropriate interactions are self-evidently required behaviours. Beyond that, some other specific professional attributes and actions called for are:

* Make their best efforts on behalf of their mentees
* Be open and honest, especially about expectations for mentoring outcomes
* Manage the mentoring process conscientiously
* Preserve the mentee's confidentiality
* Maintain professional integrity.

Effective mentoring is a skilled activity. Practitioners here, as in any other profession, have a duty to keep their own knowledge and skills up-to-date, in other words to undertake *Continuous Professional Development (CPD)*. That means that mentors should undertake training and CPD, not only for their core subject but also for mentoring itself. For professional coaches, the term 'Supervision' is used in the specialised meaning of checking one's practice, somewhat like coaching of a coach. It is therefore also advisable for mentors.

Case History 4.1 illustrates the strength of belief about the central role of people that drives leaders in the technology sector.

Various kinds of dysfunctional mentoring behaviour have already been discussed (see Section 2.4). Dysfunctional mentoring is, in fact, the result of a breach of one or more of the Mentoring Principles, a lack of skill and ineffective management. The success of mentoring depends critically on three factors:

1   Observing the mentoring principles
2   Employing specific skills to create awareness
3   Effective management of the mentoring process.

CASE HISTORY 4.1 – THE ETHICS OF INNOVATION (AND MENTORING) (TIMOTHY BRUNDLE)

In 2000 Tony Blair's Labour Government agreed with the Trade Unions to support the creation of a Defence Diversification Agency to transfer the technologies developed within the UK during the cold war into civil applications. Beating swords into ploughshares, just like in the Book of Isaiah. My mentor Ernie Shannon suggested I join them and I once more took his advice. At the time, the Ministry of Defence had some 10,000 scientists and engineers in its employment that were suddenly made available to stimulate innovation within the UK economy. They had great credentials, too. My new colleagues had invented liquid crystal displays and carbon fibre, developed satellites and sensors that had survived both space and the ocean floor. It was there that I met Antonia White, who directed the Agency's operations in Northern Ireland, Scotland and Wales.

In common with Ernie Shannon, Antonia upheld the principle that innovation was about people. Just like Apple's Jonny Ive, she argued that the technologies that were placed in people's hands had to transform their experiences for good; that you must develop empathy for those who were engaging with the technology and enable the fulfilment of their aspirations. But one of her last pieces of advice was on the subject of ethics, remarking that 'just because you can do something, doesn't mean that you should'. The way we design and use technology has an impact upon each of our lives, in good and bad ways.

EXERCISE 4.1 – HOW ARE MENTORING PRINCIPLES APPLIED IN PRACTICE?

1 A matter of choice

Imagine a company has identified a lack of skill in some technical specialisations and so it needs to increase the knowledge and skill among its professional scientists in those areas. Nonetheless, the directors do not want to recruit new scientific staff because that would increase the total wages bill. Instead, the Board instructs the Chief Executive to set up a training scheme for junior scientific staff to be supported by a group of mentors selected from the cohort of experienced employees. In consultation with others, the CEO mandates what subjects are to be covered in the training scheme, nominates the trainees and mentors and appoints a training manager to run the scheme and organise mentoring pairs. The mentors' role is to monitor their trainees' progress and help them achieve advanced qualifications in the specialised subjects. One of the trainees tells you, as his mentor, that the subjects in the course don't interest him and won't lead him in the direction in which he wants his career to go in the future. He asks for advice.

- *What advice would you give him?*
- *How should you resolve the conflict of interest between your responsibilities to the trainee and to your employer?*

(Hint: see Section 4.2)

2 What is your responsibility?

In instructing you to mentor a certain junior professional, your Chief Executive says to you, "Your job is to get him through chartership".

You therefore arrange to meet your new mentee and conduct some mentoring sessions. You tell him what subjects he should cover. You suggest exercises and direct him to journal articles and texts for private study. You make yourself available to discuss technical matters in all of the topics. However, he gives excuses for missing some of the pre-arranged mentoring sessions. During the meetings that he does attend, his responses to your questions show that he has not studied the references closely. In due course he submits his application to the learned society that is the Chartership awarding body. After due consideration the Society notifies him that his application has been deferred. An accompanying letter offers constructive advice on how a future application from him could be improved. The advice can be simply summarised as 'better preparation'. Your Chief Executive calls you in for a discussion.

- *Would the CEO be correct to blame you for the trainee's deferral?*
- *How could you have managed the mentoring engagement so that your mentee's application has a better chance of success?*

(Hint: see Principle 3, also Section 4.2)

3 What price confidentiality?

You are a senior manager mentoring a junior scientist in another department. You are aware that mentors must preserve confidentiality in order to build and maintain trust. Your mentee's line manager comes to you to ask how your mentee is progressing in training and mentoring.

- *How would you resolve the conflict of interest between the manager's need for information and the mentee's need for privacy?*
- *Your mentee has revealed personal information about a mistake he made in his previous employment. Although not illegal, the mistake if it was known would be potentially damaging to his career. Would you pass this information on to his manager?*

(Hint: see Chapter 6; also Principle 5 above)

4 Time for mentoring?

You are a senior engineer recently nominated as a mentor in a company-sponsored mentoring scheme. This new responsibility is a recognition of your expertise and senior status although it is additional to carrying out your normal work. You are conflicted between spending the time necessary to meet your own agreed performance targets and that required for mentoring.

- *What should you do to resolve the conflict?*

(Hint: see Principle 3 and 5 above; also Chapter 6)

4.2 A FRAMEWORK FOR THE MENTORING PROCESS

Traditional mentoring just happens. It is unstructured. Mentor and mentee meet at unspecified intervals to talk about their mutual professional interest in a variety of subjects. In due course, they meet less often and the mentoring relationship just fades away as one or the other moves on. This traditional mentoring arrangement has often worked well in the past and so many people will be content to leave it at that. The problem is that traditional mentoring is based on a set of unspoken expectations. Unless the mentor has a high level of interpersonal skill, the traditional approach carries a high risk of wasted time and dysfunctional outcomes.

Traditional mentoring, as an unstructured, open-ended commitment, can seem like large and nebulous demand, one that runs the risk of rambling on interminably. Many inexperienced mentors fear that if they give their mentees completely unrestricted access, they will drop in whenever they feel the need and soak up huge amounts of the mentor's time. The best way to avoid problems of unstructured mentoring is to be aware of the principles discussed in the previous section and to establish ground rules about how the mentoring process should work in practice. Modern mentoring is based on a set of clear and explicitly agreed procedures and a systematic approach to creating an outcome.

Mentoring relationships can cover a very wide range of issues. It helps to narrow it down and think of the aim for a mentoring engagement as a specified piece of learning or a problem to be solved.

For example, the aim might be to acquire a particular technical skill, to achieve a career transition, to gain a higher qualification or to address a motivational issue. Whatever the aim is, a mentoring engagement can range from a single short session up to a series of sessions spread over a period. A larger mentoring engagement with a broad aim can be divided up into a series of lesser objectives with steps leading to the desired outcome. Whatever the aims are, within each mentoring engagement there are four steps:

1   Connecting
2   Scoping
3   Mentoring sessions
4   Review and close

4.2.1 Step 1 – Connecting

Connecting with a mentor is the first and biggest challenge for younger professionals who would like to receive mentoring. In the context of traditional or informal mentoring, connecting usually starts with a coincidental meeting and a conversation which leads to a series of *ad hoc* meetings. In the context of a mentoring scheme, connecting is managed by whatever arrangements for pairing are established within the scheme. Such schemes range from highly structured to quite informal depending on whether the responsibility for the initial connection is taken by an organisation or by individuals.

Informal mentoring schemes are offered by some organisations, especially professional associations and institutions. They are, in effect, simple contact services. Typically, the association will maintain a register of self-nominated mentors along with some personal information including their areas of expertise and contact details. Association members who are seeking a mentor can examine the register and initiate the contacts themselves, with or without some intervening filtering process. Other than providing general advice and maybe some training, the institution takes no further part in individual mentoring arrangements.

Formal mentoring schemes are relatively common nowadays within large scientific and engineering companies and public sector organisations. Typically, a central organiser oversees the scheme, recruits a panel of mentors and gathers potential mentees. Connecting is carried out through matching processes of varying complexity. They can range from informal and *ad hoc* arrangements to more sophisticated processes involving personality profiling questionnaires that are intended to allow matching of compatible personality types, aptitudes and professional interests. The matched pair then undertakes mentoring meetings. The number and frequency of meetings may be predetermined. Various requirements for recording, reviewing, reporting &/or evaluating progress may be specified. In the best schemes, extensive mentor training is provided and arrangements made for professional supervision of mentors' performance and progress. The structure of a mentoring scheme is outlined in Chapter 6 below.

As a senior professional volunteering to be a mentor, you may facilitate connecting but should not initiate the contacting process. In particular, you must not pressure anyone to be your mentee or try to 'sell' your mentoring services. What you can do is to make yourself approachable generally and, if the occasion is appropriate, talk to potential mentees about how mentoring works or arrange to meet and chat at another time. If they express confusion or ignorance about how mentoring works then you can offer brief general information about principles, what mentoring is and what it can and can't do. Explain why they must take the initiative to approach a potential mentor. Suggest that potential mentees look for mentors by attending professional meetings, using social media or signing up to an informal scheme. If they show interest and if you think you may be able to help, say that, in principle, you may be willing to help and that you also know others who may be able to help. Offer guidance about the best way to approach a potential mentor in a scoping meeting (next section)

4.2.2 Step 2 – Scoping

Before agreeing to mentor anybody you first need to see if you might be a good match with the potential mentee, that is if you could work together productively. The best way to approach this stage is to hold a brief scoping meeting. In coaching jargon, this is a 'chemistry meeting', a term derived from 'personal chemistry', referring to your ability to get along together. As well as finding out what the prospective mentee wishes to achieve in mentoring you can also find out what they already know about the process. If that is only very little or even misguided you can explain mentoring principles and processes in outline and set the ground rules, especially mentioning self-responsibility, accountability and confidentiality. In what is effectively a 'taster' session for the mentee, you should use the skills described in the last chapter as well as the principles and processes described in this chapter. Exercise 4.2 gives some suggestions for ways to approach the meeting.

Some basic expectations of *an acceptable mentee* could include:

Self-responsibility and intrinsic motivation – As an essential condition for working together, ask that they take responsibility for their own development, are proactive in pursuing their own goals and accept responsibility for the outcomes. You are not a school teacher. Your job as their mentor is strictly limited to facilitating their development process.

Openness and honesty – In mentoring, it is normal to uncover sensitive, personal problems that can block your mentee's progress – family or business problems, unreasonable expectations, anxiety and lack of confidence. Problems can only be resolved if your mentee has the courage to be open and honest. In asking for openness and honesty, you need to develop trust, show empathy and maintain confidentiality. You must be open and honest about yourself, your own experience, background and values.

Equal negotiation and accountability – As an adult learner your mentee has to feel in control of his/her own learning process. The self-responsibility principle means that your mentee is free to disagree with anything you propose, such as exercises and reading to be completed between mentoring sessions. Depending on the subject to be addressed, you may need to negotiate the conditions under which you can hold the mentee to account for agreements and performance.

Diligence and respect – You are volunteering your time and expertise for the benefit of the mentee. In return, you can reasonably ask for and expect the basic forms of acknowledgement and respect – punctuality, courtesy, self-responsibility and diligence. Your mentee is equally entitled to reciprocal behaviour from you.

If (and only if) you are satisfied, then tentatively propose a short series of mentoring meetings to address a specific topic as described in the next section. Ask that your potential mentee gives you a clear decision to accept or decline, adding that you do not require any explanations. Allow a few days 'cooling off time'; the decision can be given by e-mail or phone.

When scoped in this way it is unlikely that either person will conclude that they should *not* work together. But it can happen. *Some reasons for declining* to mentor a particular individual include:

- A conflict of interest and/or time constraints – by far the most common reason

- The mentee has made inaccurate assumptions about your experience and what you can do
- The mentee's approach and lack of experience suggest that they have unrealistic expectations
- You suspect he/she is 'box-ticking' with no real intention to engage fully in mentoring
- You suspect he/she is not being open and honest about something important.

If your conclusion is unconditionally negative, then it is important to decline without further ado. In declining you should avoid giving gratuitous offence. After all, your suspicions might be incorrect. Let them down politely, *e.g.* "I don't think I'm the right person to mentor you". If it is appropriate refer them on to someone who could be a better fit. Remember that the potential mentee has a reciprocal right to decline to accept you as a mentor.

Defer the decision if you're not sure but take no more than a few days to think about it. Aim to get clear in your own mind about what is causing your hesitation. Ask if you can consult with mutual acquaintances who have worked with the mentee. Then make a clear and unambiguous decision within an agreed period of time.

Confidentiality is an especially important aspect of mentoring in that it is the main way that trust is established. In formal mentoring schemes sponsoring organisations have a legitimate interest in the outcomes of mentoring. However they should not be party to confidential details of what happens during mentoring sessions. The best way round this apparent paradox is to hold an initial three-way meeting between the mentee, yourself and the sponsoring line manager. In the three-way meeting the mentee should say what outcomes he/she is aiming for and together the meeting should agree what indicators would be acceptable evidence of progress. It is not necessary for the line manager to know how the progress was achieved. A 'black box' approach to mentoring preserves confidentiality.

Referring on is the process of directing a mentee to someone else. A rare but significant circumstance is that, unknown to the mentor, the mentee maybe suffering from mental illness such as depression, or be the victim of maltreatment, say bullying or domestic abuse. Any attempt at intervention by you as an unqualified person, however well-intentioned, risks exacerbating the problem. In these circumstances mentors have an obligation in the mentee's best interest to refer their mentee to appropriate professional support. However, referring on can be tricky. A mentor cannot break confidentiality by unilaterally telling a third party about his/her concerns. In scoping you should prepare for this eventuality (even if it is unlikely) and plant a virtual marker flag by saying something like, "If a circumstance should arise during our work together which is outside my professional competence, I will let you know and offer to help you find more effective support than I can provide".

EXERCISE 4.2 – SUGGESTIONS FOR AN INITIAL SCOPING MEETING

Either role play with a partner **or** try out the following when you have a new mentoring engagement ahead.
Make a date and time for a 30-minute meeting with a prospective mentee and proceed as follows:

- Arrange to meet somewhere neutral and public so that the prospective mentee can feel safe and comfortable (e.g.: a café or hotel lobby)
- Be relaxed, informal and aim to put the prospective mentee at ease. He/she will expect you to lead. Help to build rapport with a few minutes (only) of normal 'get-to-know-you' social chat – ordering coffee, talk about people and places you both know and topics of mutual interest.
- Begin by saying what the purposes of the meeting are and how long you propose to take (about 30 minutes is normal). Clarify that there is no obligation for you to work together in mentoring and no explanations for declining are needed.
- Listening is far more important than talking. It is essential that the potential mentee senses they have been heard and understood. If you catch yourself talking too much, ask questions, encourage the prospective mentee to talk and listen actively.
- Explore what the mentee thinks you might be able to do for him/her. Question your own assumptions about what you think he/she might need.
- Explain the principles of mentoring (briefly), especially your own role and its limitations, confidentiality and referring on. Outline the normal frequency and timing of meetings.

- Explain what you would expect of a mentee including accepting responsibility, openness, honesty, commitment and diligence.
- Examine any possible constraints on availability, scheduling and timing.
- Look out for possible conflicts of interest and explore them carefully.
- Notice your own feelings and reactions; watch for matching/mismatching and incongruence.
- Think carefully about whether or not you really can provide what the mentee needs and decide whether or not you want to work together.

4.2.3 Step 3 – Mentoring sessions

In a successful scoping meeting you will have made arrangements for a series of mentoring sessions. In every session the mentee provides the topic and substance of each meeting. Your primary role is to structure the process, to facilitate the mentee's thinking and to keep the mentee focussed on the agreed outcome. In order to settle worries about open-ended commitments, wasting time and unspoken expectations, pay attention to the following:

Preparation – Before the first meeting it will help both of you to have an idea of the mentoring objective and some basic idea of how your mentee approaches learning and problem-solving. If you ask your mentee to do some preparation before they come to the meeting, they will arrive in a more receptive frame of mind. The simplest way is to ask your mentee to reflect on what they want to achieve and then to complete a brief self-assessment such as the one in Exercise 4.3. Some mentoring schemes use one of the numerous psychometric profiling questionnaires, although it is not absolutely necessary. Even though they require expert administration, the accuracy of such profiling techniques is often questioned. Mentors may find psychometric profiling to be useful not so much for their accuracy but rather they provide an in-depth investigation of the mentee's characteristic thinking styles which can help to raise their self-awareness.

Scheduling – Propose a limited series of meetings over a definite period. Even a large or complex topic can be sub-divided into stages. For each stage, a series of three or four meetings would be normal, each 60 or 90 minutes long, spaced two to four weeks apart with private study in between sessions. Conclude with a review of progress at the end of each session and the conclusion of a whole series. If the review finds that the topic has not been fully addressed then negotiate another series as required. Structuring in this way takes advantage of our unconscious tendencies around time boundaries. If a mentee knows that time is almost up they will often spontaneously arrive at a conclusion.

Location – Although a mentoring session appears relaxed and informal, it is actually closely focussed and structured. It is important to arrange the immediate environment to support clear and continuous thinking. Arrange to meet in a place where you can talk in private without being overheard. Similarly, turn off phones and make whatever arrangements you can to avoid interruptions. A neutral space is preferable. Meeting in your office or the mentee's office is usually not ideal because the environment brings with it all kinds of unconscious signals that can interfere with thinking. Worries about unequal status or uncompleted work, for example, can be distracting or distorting. It is also more likely that you will be interrupted. Many organisations have meeting rooms which can be booked for specific periods. Phone and internet video meeting platforms can also be used successfully.

EXERCISE 4.3 – PREPARATION FOR MENTORING

You can offer the questionnaire below to your new mentees. It will help them to think through what they would like to achieve in mentoring with you. The information they provide will help you to help them.

Also, try it yourself. You will raise your own self-awareness and understanding of what you are asking your mentees to do.

Take some time to reflect and to answer the following questions:

What issue do I want to address in mentoring? Be specific. You can enlarge or change it later.
What would be my ideal outcome?
What are my strengths?

What am I prepared to be held accountable for?
What are my most important values?
What are my signature strengths? (*i.e.:* strengths that are especially well developed)

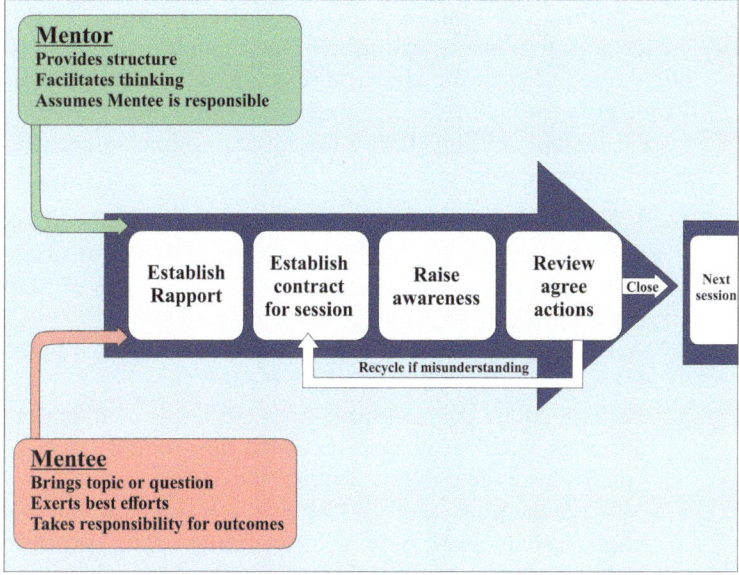

Figure 4.1 Diagram of a mentoring session.

A mentoring session consists of four stages (see Figure 4.1) as follows:

1   Establishing rapport
2   Contracting
3   Raising awareness, and,
4   Reviewing

1   ***Establishing rapport*** – Since you have met before either in the scoping meeting or in a previous session, you can begin with a brief greeting, re-connecting and ensuring that your mentee is comfortable and ready to start work. These so-called set inductions are important for psychological comfort and subsequent interaction. In some societies set induction rituals are much more

elaborate and formal than in others. Look out of signs of physical or psychological discomfort. A good place to start is to review progress on tasks agreed at the last meeting. Whatever precursor activities you do, take care that they do not occupy any more than a few minutes of the total time agreed. One of your responsibilities as in mentoring is managing the time for the session.

2   *Contracting* – The term refers to agreeing a psychological contract and not any kind of formal or written contract. In this sense, contracting is a negotiation around an informal but explicit agreement about what you're going to do in mentoring. Its purpose is to manage expectations. ***Contracting is essential*** for a successful session. Take a few minutes immediately after establishing rapport to:

1   Enquire about the subject that the mentee wants to address in the meeting, *e.g.* a technical subject or a job transition. This is an essential application of the principles of Self-responsibility and Intrinsic motivation.

2   Ask about the mentee's expectation of you for this meeting. Clarify what you can and can't do.

3   Explain that your conversation will be confidential, although with prior agreement the need could arise for you to share the outcomes with a third person.

4   Manage the time carefully and precisely. Explain that you might have to seem rude by interrupting a long speech in order to manage the time better. Say what time this meeting will end and stick to it.

5   Explain your expectations of mentees in general and ask for understanding and agreement.

6   When you have experience of working with an individual mentee it may be helpful to ask how he /she would like you to conduct the session. That could range from being supportive and sympathetic at one extreme to giving constructive feedback and offering challenge at the other.

7   Ask if the mentee wants you to hold them to account for agreements and progress they make. This will be essential for giving structured feedback on performance later on.

3   ***Raising awareness*** – After contracting the dialogue moves on to the substantive business of the mentoring session, the part that takes ***the great majority of the allotted time***. Since all mentoring is outcome-orientated, the general pattern starts with raising the

topic, moves through exploration, ideally to arrive at an insight or discovery that leads to an action or a decision. Your objective is to facilitate the mentee's thinking so that they can address the subject they raised during the contracting stage using their own personal thinking processes. In the session generally, but especially in this phase, you will simultaneously be using all five of the essential skills described in the Chapter 3. *You will be listening much more than talking.* Any talking you do will be in the form of questioning sequences and reflective listening that are essential to facilitate the mentee's thinking. You not only listen actively and respond, you also need to manage yourself and the session efficiently and unobtrusively. Wherever possible avoid giving direct advice. If you have to give feedback, make sure you first make the offer and, if accepted, then respectfully deliver and properly structure the feedback as described in the previous chapter.

It usually helps to follow one of the *coaching models*, that is frameworks for managing a structured conversation. These models are questioning sequences of reflective listening that move through stages intended to arrive at the sought-after conclusion. Three such well-established coaching models are known by their acronyms:

GROW – Goals, Realities, Options and Way forward. Brought to prominence by Sir John Whitmore from earlier work in the 1980s, this is the best-known and most used model (Whitmore, 2002).

CLEAR – Contracting, Listening, Exploring, Action and Review (Hawkins & Smith, 2006)

FACTS – Feedback, Accountability, Courageous goals, Tension, Systems thinking (Blakey & Day, 2012).

In using one of these frameworks do not announce explicitly what you are doing or try to be too rigid in structuring the session. Over-awareness of the structure and your facilitative processes will result in feelings of awkwardness and self-consciousness that distract the mentee. The ideal is a feeling of informality along with psychological safety for the mentee. Some mentors say it is like *dancing in the moment*, a metaphor that expresses fluidity, flexibility and partnership in responding to the mentee's needs in the context of an underlying structure and clear intent. Exercise 4.4 gives an illustration of how the best-known model can be used. Section 5.3 describes specific application of the GROW model in technical mentoring.

4   *Reviewing* – About five minutes before the scheduled end of the session, quietly mention the time and draw the substantive discussion to a close. Before closing, take a few minutes for the following:

Ask the mentee to review the salient facts, conclusions and actions. Check for agreement. If the mentee has followed his/her own thinking process rather than yours, then they will 'own' the outcome, their commitment will be stronger and success is more likely.

Propose private study to be completed as preparation for the next meeting. In fact, the greater part of the mentee's own work and development is carried out between mentoring sessions. Private study tasks will be related to what you have been discussing and what you think will help the mentee to understand the subject and prepare for the next session, *e.g.*: taking an agreed action or gathering information such as reading a technical paper or meeting someone to gain some information. Some mentors call this private study or preparation 'homework', although for others the word has unpleasant connotations. Whatever you call it, negotiate it fairly and do not impose your own demands. The point is that by asking the mentee to report back to you in the next session you tend to make fulfilment more likely. In support of the private study and as a record of progress it often helps to suggest that the mentee keeps a private journal to record their reflections and progress. Make it clear that you will not be asking to see it. The journal is a private document.

Finally, ask if the mentee wants to meet for another session. If so, *make arrangements for the next meeting* accordingly. Agree the date, time and place.

A note about taking notes: Professional scientists and engineers are accustomed to taking notes during meetings. Indeed, for many people note-taking is the automatic default in any meeting and their notes may feel to them like a security blanket. In a mentoring session, however, taking notes can be distracting for both parties. To write notes you have to look away, break rapport and slow the pace of interaction, all of which distracts the mentee's thinking process. It is much more important for you to listen actively. If you feel that you need notes, then write up them up *after* the meeting.

It is important to remember that your notes are confidential and personal. Your mentee has a legal right to see them and to complain if he/she feels misrepresented in any way. To protect both yourself and your mentee it is better not to include detailed reports of conversations. As an *aide memoire* for use in future mentoring sessions, note only keywords about significant issues, conclusions and agreements. It is sensible to tell your mentee what is and what is not included in the notes. When the whole mentoring engagement ends, offer the notes to the mentee or else destroy them and make it clear that you have done so.

Your mentee may also want to take notes but, again, this can be distracting for his/her own thinking processes. As an alternative, you can say that there will be review at the end of the session in which they can take notes. Another alternative is to use mobile phones for recording. Whatever either of you proposes in this regard, the important factors to consider are openness, mutual agreement, preserving the mentee's confidentiality and, most importantly, the extent to which it causes a distraction. It depends very much on what works best for the individual.

4.2.4 Step 4 – Review and close

Closure as the final step is the compliment of set induction at the beginning and is just as important psychologically. Its significance lies in embedding the mentee's change and allowing smooth onward movement. The structure of individual mentoring sessions is a small-scale version of the structure that comprises the whole series for the mentoring engagement. At the end of the final mentoring session, take five minutes or so to review overall progress towards the aim that was agreed at the start of the series. This will create clarity for both of you about how far you have come and how much remains to be done.

Ideally, the mentee will have resolved the problems or gained the knowledge and understanding that was the objective of the series of mentoring meetings. Give acknowledgement and positive feedback as appropriate (Section 3.6). Celebrate success, acknowledge and mark progress. Perhaps another topic for mentoring will naturally present itself. Provided that you are prepared to continue working with this particular mentee, check they also want to continue working with you. If so, negotiate and plan a further series of mentoring meetings. Follow exactly the same structure and process and start a new series.

Conversely, if the objective has not been achieved, then think about whether or not further mentoring with you has a chance of achieving the outcome and, if so, how. If you don't think you can help further then gracefully acknowledge the mentee's efforts, outline your reasons, close and move on. If appropriate, suggest other sources of help and refer on. Whether or not the mentoring has achieved its objective, the mentee must make an active choice to continue working or to move on. If he/she does not wish for further mentoring from you, then gracefully agree to close the engagement. Don't argue or try to 'sell' further mentoring. The Self-responsibility Principle means that it really must be his/her decision. Acknowledge the mentee's efforts and contributions to the process. Asking for your mentee's feedback on the mentoring will help you to improve your own practice.

If mentoring is within an organisational context, then agreed feedback to line management and/or the HR department may be necessary. The best way to achieve this is in a closing three-way meeting. It is parallel to the initial three-way meeting between yourself the mentee and an appropriate manager. Make sure beforehand you have agreed with the mentee what may be said and what to keep confidential. Generally, anything revealed within mentoring session should be kept confidential (Figure 4.2).

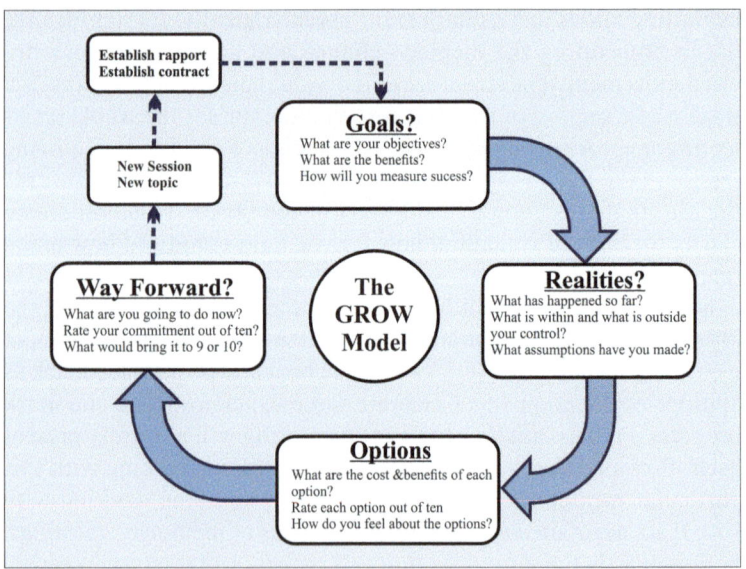

Figure 4.2 The GROW model for mentoring.

EXERCISE 4.4 – THE GROW MODEL IN ACTION

Integrate this exercise into your next mentoring session. Applications of the GROW model specifically for technical mentoring and career transitions are described in the next chapter (see Chapter 5 Mentoring in Practice). A period of one hour is normal, although useful work can often be achieved in 10–15 minutes.

Establish rapport and contract for the session as described above.

Then follow these four steps:

1 *Goals* – Begin by asking about the topic and your mentee's objectives for this session. Avoid jumping to conclusions. Ask concise questions, summarise, make observations and check your mutual understanding of the issue. Watch for non-verbal signals – if you see something you think is relevant, don't assume what it means – ask. Some relevant questions might be:
   - *So, I understand that you want to think about X. Please can you give me more detail?*
   - *What would be your best outcome in this situation?*
   - *What is the decision you need to make?*
   - *When this session comes to an end what would you like to have achieved?*

2 *Realities – Encourage your mentee to explore the facts and realities of the situation as distinct from focussing on his/her wishes and fears. Help them distinguish fact and evidence from interpretation.*
   - *What has happened so far?*
   - *What can you control in this situation and what is outside your control?*
   - *What is stopping you from doing that?*
   - *What assumptions have you made?*

3 *Options* – Explore the options and actions open to your mentee and their preferences. Avoid the temptation to impose your own answer or to give direct advice or opinion. Always assume that the mentee is the best source of the solution. Remember to examine feelings as well as facts. Focus on desired outcomes and

solutions rather than the painful or difficult detail of the problems. If the best option seems reasonable and doable but your mentee is still hesitant, bring that to his/her awareness. Invite him/her to find out what's really happening.

- *What are the benefits and costs of each option?*
- *What advice would you give a friend in this situation?*
- *Rate each choice out of ten.*
- *What is your intuition telling you?*

4 **Way forward** – Encourage your mentee to identify a conclusion or make a decision. Never try to impose your preferred solution. It is essential that they 'own' the conclusion. Make it explicit. After arriving at a conclusion or decision, go on to encourage and support the mentee to take action and exercise initiative. Don't offer to take over or do it for them. Check the duration, scheduling, ability and other conditions and offer to hold the mentee to account. Acknowledge and give genuine and appropriate positive feedback.

- *So, what are you now committed to achieving?*
- *On a scale of 1 = no commitment to 10 = full commitment, Where are you now?*
- *What would make it a 9 or 10?*
- *What now are the steps from here?*
- *How will you measure success?*

Note: You do not need to follow this scheme rigidly. Allow the mentee to change their mind as new insights occur. Be flexible, go back to an earlier phase and re-iterate as needed.

Chapter 5

Mentoring in practice

5.1 TYPICAL MENTORING CHALLENGES

The preceding chapters give an account of mentoring in general. Even though every mentoring relationship is unique there are some generic mentoring situations found in all professional science, engineering and technology practice. This section describes and offers advice about some of the most common generic mentoring situations:

1 Establishing the baseline – finding out your mentee's knowledge skills and learning style
2 Technical mentoring and professional skill training – on-the-job mentoring skills
3 Mentoring for professional qualifications – helping your mentee get further qualifications
4 Professional attitudes and motivational interviewing – when motivation is a problem
5 Career transitions – helping your mentee make the next career move
6 Self-confidence and low self-esteem – helping with a very common, but usually hidden problem
7 Intercultural mentoring – being a mentor in the wider world.

5.2 ESTABLISHING THE BASELINE

Good mentoring requires a clear understanding of our mentees as individuals. Dysfunctional technical mentoring often happens when we wrongly assume that we already know what our mentees know. Acting on the assumption that your mentee has a higher level of knowledge than they actually do possess will mean that your explanation leaves them feeling confused and ignorant. Alternatively, if you assume they have a lower level of knowledge, you not only waste time but also appear patronising. That damages the professional relationship while not answering the mentees' real underlying questions. Sometimes you will meet mentees who seem to know a great deal in theory but falter when they face real-world practical situations. Another cause of mentoring failure is the tacit assumption that your mentees learn and process information in exactly the same way as you do. No matter what issue your mentee brings for mentoring, you always need to begin by finding out where the individual in front of you is starting from. Even a little basic understanding will help. Before engaging in any technical mentoring with a mentee it will help to find out:

a   How much does he/she about the subject right now? – assessing knowledge level
b   How well does he/she use the knowledge they already have? – assessing cognitive skill level
c   In what way does he/she best learn? – assessing preferred learning style

5.2.1 Assessing knowledge level

Educators say that there are *three domains of learning – Knowledge, Skills, and Attitudes (the KSA's).* In 1956, a group of educationalists published a simple method of general assessment in each of these three domains. *Bloom's Taxonomy of Educational Objectives* defined standard levels of learning (or a taxonomy) in each domain. Krathwohl (2002) gives an overview of the revision of Bloom's taxonomy by Anderson *et al.* (2001). This approach is based on the simple observation that learning progresses in a small number of generic stages specific to each domain and that we must complete each level of learning before moving up to the next. Therefore, we can identify any mentee's generic level of knowledge by asking only few well-chosen technical questions. The method is not wholly reliable because everyone has different knowledge gaps whatever the level they are at, but it does give a useful way of finding out roughly what your mentee knows before you launch into an explanation.

Bloom's Taxonomy in the Knowledge Domain defines four levels of knowledge that apply generically to any subject. The most basic level is factual knowledge, moving up in turn to conceptual, procedural and metacognitive knowledge at the highest level. The mentee's response to a few straightforward technical questions reveals where the mentee's current level of knowledge is on this scale. The answer indicates the level of knowledge at which you can begin instruction and discussion. Details are given in Table 5.1 (below) with an example in hydrogeology.

Table 5.1 Bloom's hierarchy of knowledge levels

Level of knowledge	Meaning	Examples in hydrogeology
1. Factual knowledge	The essential facts and ideas required to understand a basic technical discussion and solve problems in the field, including essential terminology, definitions and elements	• Terminology of elementary hydrogeology • Groundwater flow and wells • The hydrological cycle • Basic field geology
2. Conceptual knowledge	Knowledge of how the basic elements fit within larger structure and how the models relate them to each other, including classification operating principles and common principles and theories	• Darcy's Law • Basic hydrochemistry • Groundwater in relation to local geology
3. Procedural knowledge	Knowledge of skills needed and ways of thinking; techniques and methods employed; how investigations are carried out; instruments and sources of information, when methods are appropriate and their limits	• Well construction & testing • Groundwater flow modelling • Mass balance calculations • National policies & legislation
4. Metacognitive knowledge	Strategic knowledge, usually with long experience; ability to be creative and develop new models in the field; understanding of the thinking skills and self-awareness	• Effective solution of complex local problems in groundwater studies • Ability to teach hydrogeology effectively

For example, imagine that a senior hydrogeologist takes a junior environmental scientist for some on-the-job training at a location in the field. The senior hydrogeologist brings out a detailed geological map of the area and asks, "Where and how would you construct a groundwater well in this area?" This question tests for level 3 – Procedural knowledge (see Table 5.1). If the environmental scientist's answer to the first question is correct, then the next question tests for the advanced level (level 4 – metacognitive). Alternatively, if the answer is incorrect then the next question moves down the hierarchy to establish the whether the junior's generic level of knowledge is very basic (level 1 – factual) or somewhere in between (level 2 – conceptual). The subsequent discussion starts at this level and, ideally, moves on into areas that he/she did not know to begin with.

5.2.2 Assessing cognitive skill level

We can make a useful practical distinction between how much we know and how well we use what we do know; in other words between our level of knowledge and our level of cognitive skill. Professional institutions define the levels of competency or skill required for professional status. The dictionary definition of competency is, "The ability to do something successfully or efficiently". Skill is simply a synonym, "The ability to do something well; expertise" (both definitions from the Oxford English Dictionary, 1964). Therefore, *competency or skill is knowledge demonstrated in practice*.

According to Bloom's Taxonomy of Educational Objectives in the Cognitive Domain, we learn by successively mastering each of six levels of skill (Krathwohl, 2002). From the most basic level in order upwards, they are: remembering, understanding, applying, analysing, evaluating and creating (see Figure 5.1). Remembering is the most basic cognitive skill because understanding is not possible unless we can recall information. Similarly, unless we understand we cannot fully apply a skill. The ability to apply a skill is necessary to allow analysis which, in turn, is necessary before evaluating and creating are possible at the highest levels of thinking skill. As with assessing knowledge levels, asking a few simple questions allows you to identify the generic level of cognitive skill your mentee has achieved. Having a rough idea of your mentee's cognitive skill level will help you decide how you can best help him or her to improve their skill in some particular area.

Table 5.2 Hierarchy of cognitive skills in Bloom's Taxonomy, illustrates examples of applying the hierarchy within the field of mineral exploration. So if, as a senior exploration geologist you are

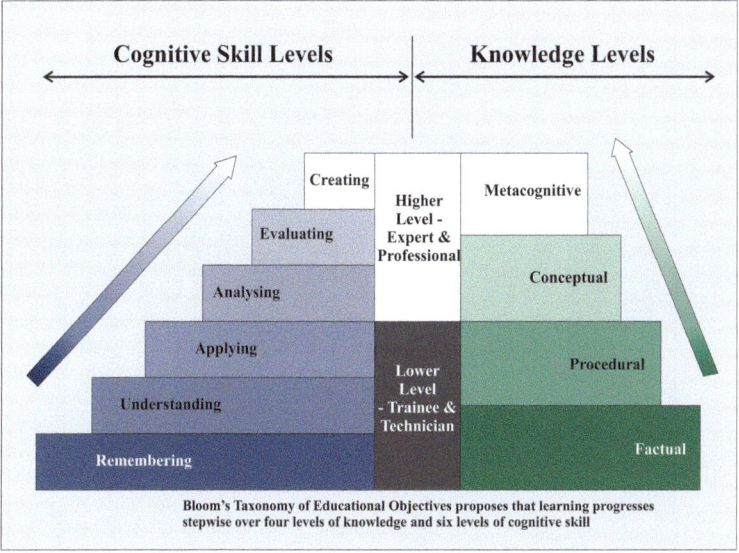

Figure 5.1 The hierarchy of cognitive skill levels according to Bloom's taxonomy.

asked to help a new geologist arriving on site, you might establish his cognitive skill level by asking a few questions indicated by the levels in the right-hand column. The same principle is easily transferred to any other scientific or engineering application.

The distinction between higher and lower levels of cognitive skills is important in understanding the level of professional competency attained. *Lower level cognitive skills* (remembering, understanding and applying) confer the basic ability to manage data. They include applying a pre-defined set of rules to an operation or investigation. Competently performing standard operating procedures such as those carried out by trained technicians and computers is typical of lower-level cognitive skills, *e.g.* recording of laboratory observations and samples *pro forma*. *Higher level cognitive skills* (analysing, evaluating and creating) are learned through advanced training, practice and experience. They usually employ intuition, short cuts and rules-of-thumb (heuristics) along with further study and application, *e.g.* analysing sample results, evaluating their significance for the investigation as a whole and creating new interpretations. *The ability to employ higher level cognitive skills is a distinguishing feature of professional competency in technical fields.*

Table 5.2 Hierarchy of cognitive skills in Bloom's taxonomy

Level of cognitive ability		General description	Examples in mineral exploration
Lower – technician and learner	1. Remembers	Can recall instructions and information given previously	Knows the routines for sampling drill core, including quality control checks
	2. Understands	Can understand and re-phrase instructions and information given	Can explain how and why a sampling programme works
	3. Applies	Can apply already known information to a new situation, or instructions to new observations	Given targets, can design an exploration drilling programme
Higher – professional and expert	4. Analyses	Can analyse information into components and their structure	Given exploration reports and maps of a new area, can separate facts from opinions, classify the data collected and see what further work is needed
	5. Evaluates	Can assess the value of a set of observations or ideas in the circumstances under consideration	Given exploration reports and geological maps for an area, can make a judgement, backed by sound reasoning, about whether or not to proceed with exploration
	6. Creates	Can create a new idea or structure from available information; gives new meaning to existing models	Can assemble available observations into an ore deposit model which explains all the exploration data collected so far

5.2.3 Identifying a preferred learning style

The knowledge we possess and the skill with which we use it is uniquely structured in our individual consciousness. In giving explanations, we tend to use the thought patterns and circumstances in which we personally learned to understand the material we are trying to explain. When someone fails to understand our explanation it is often because of differences in our *preferred learning styles.* Knowing your mentee's preferred learning style allows you to adapt the manner and the context in which you explain any technical subject.

In 1984, Kolb published the *Theory of Experiential Learning.* He proposed that learning is a process in which knowledge is created through the transformation of experience. The process consists of a cycle of four stages – experiencing, interpreting, generalising and applying. Experiential learning theory was developed further by others, most notably Honey and Mumford (1992) who proposed that Kolb's four stages were characteristic of four distinct learning styles (see Figure 5.2 Honey and Mumford's learning styles).

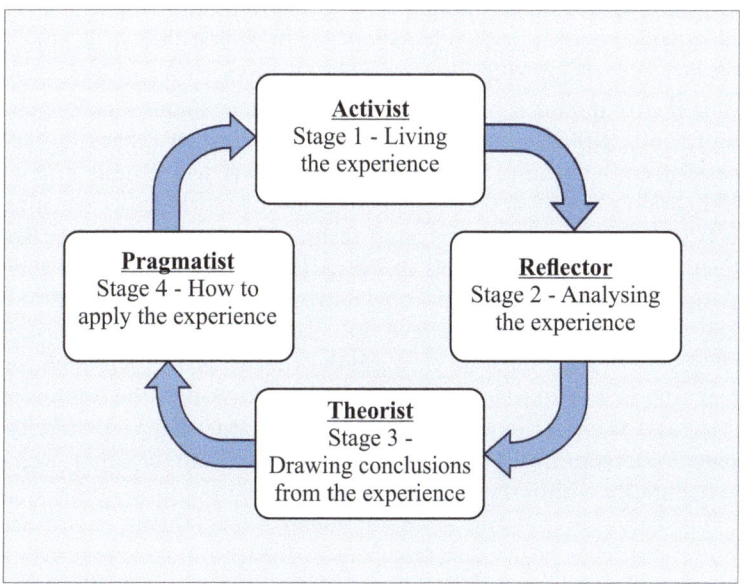

Figure 5.2 Honey and Mumford's learning styles.

According to Honey and Mumford, each mentee should show the characteristics of one of four types of learner:

1 *Activists* – People who seek challenge and immediate experience. Open-minded and gregarious, they tend to involve themselves with the learning activity first and think about the implications afterwards.
2 *Reflectors* – People who stand back, gather data and analyse experiences. They tend to be thoughtful and cautious before reaching a conclusion.
3 *Theorists* – People who think things through in logical steps and reject subjectivity. They tend to gather information systematically and use it to attempt to develop a coherent theory about an experience.
4 *Pragmatists* – People who are practical and enjoy decision-making. They like to apply theories and investigate how theory works in practice.

Learning theory and its applications are detailed in Beard and Wilson (2013).

Although these styles are simplified characterisations, individuals tend to show one dominant style along with subordinate aspects of others. Questionnaires are available that reveal an individual's preferred learning style and subordinate styles. It is easier, if less reliable, to find an approximation by asking the mentee to describe a previous learning experience that was for him/her especially successful, enjoyable and rewarding. Their answer will usually give an indication about whether the person's best experience of learning was mediated primarily through feeling (Activist), watching (Reflector), thinking (Theorist) or doing (Pragmatist). You can then design your instruction to copy that experience as far as possible. Education theorists believe these styles are also the necessary stages that we all must go through in order to properly embed the learning in our individual consciousness. Although the dominant style and stage is where instruction should start from, for a complete understanding it will be necessary to build on the other stages sequentially within the learning programme.

For example, imagine a senior scientist specialising in applied botany who has been asked to supervise the training and mentoring for a young graduate. When questioned about an enjoyable learning experience the mentee responds that she really loved the undergraduate field course. Asked to give more detail, she enthusiastically describes not only the flora of her field area but also the landscape and its ecology. She greatly enjoyed the vigorous outdoor work and the feeling of independence. From this, it looks as if she is an Activist, so the senior botanist organises the introductory part of her training around independent field work, starting by sending her off on her own to collect some primary data. Subsequent stages of her learning will be discussion and reflection on the field observations, creating hypotheses about the ecology, soil conditions, *etc.* and then planning the next steps of the investigation. Comparable variations would apply to mentees with other learning styles.

5.3 TECHNICAL MENTORING AND PROFESSIONAL SKILL TRAINING

Professional skills are exercised in the laboratory, in the field, on a construction site, in a clinical setting and wherever else scientific, engineering and technology practitioners work. An individual mentee's professional competency often becomes evident only later in meetings, conferences, discussions and in written reports. Here are some common situations you might meet at work:

- A younger colleague comes to you and asks, "How do we set up the equipment for this investigation?", or "How do you know when to use this technique?" What is the best way to respond?
- A new graduate is recruited to work in your laboratory. You have to show her around, explain what your team does and integrate her into the work of the laboratory. What is the best way of accomplishing this task?
- As a training supervisor, you have been given the responsibility to see that that a trainee satisfactorily covers a CPD module within your organisation's training programme. How should you tackle that responsibility?

Addressing technical questions is the most common place for a mentoring engagement to begin. Indeed, some people make the mistake of thinking that technical instruction is the whole of mentoring. As we have seen, there is much more to it. Technical mentoring can simply be passing on some specific knowledge but more commonly it involves some element of cognitive or manipulative skill training as well. In most professions skill training is more than just showing someone how to use a piece of equipment or the best way to carry out an investigative routine. It nearly always includes developing the cognitive skills characteristic of the profession in question thereby facilitating the mentee's move from lower levels to expert levels. For example, the skill underlying diagnosis in medical science is more than just knowledge of the symptoms and aetiology alone, it is a characteristic way of thinking that uses a set of heuristics that need to be experienced and learned.

At the beginning of a technical mentoring engagement it will help both you and your mentee to clarify exactly what your respective roles are in this situation. The mentee's role is to take responsibility for progressing his/her own learning. As a mentor, your role is to *facilitate* learning but not necessarily to carry out the instruction yourself. Of course, mentors often do have to give some technical explanations and demonstrate techniques in order to help a mentee. The difference between a mentor and an instructor is similar to the difference between a tutor and a lecturer in a university undergraduate course. Even though tutoring and lecturing are often carried out by the same person in a particular subject, the roles are actually separate and distinct, one is advisory while the other instructional (see above, Section 2.2).

We have already seen how self-motivation is an essential condition for a mentee's professional development. Rather than delivering a load of technical information yourself, it is always more effective to help your mentees to acquire the knowledge for themselves (and less work for you). The most effective procedure is an adaptation of the classical university tutorial in which the tutor sets the students a task or a problem and asks them to find their own solution. Typically, the tutor directs the students to relevant sources of technical information – journals, papers, texts, *etc.* – and then asks them to present their conclusions at a future tutorial meeting. At the start of the tutorial session the tutor may frame the topic by giving a short talk around generalities and principles. Then he/she presents the allocated task or problem and the tutorial group engages in discussion. Depending on the length of time and the depth of investigation, one or more students may be asked to present a paper detailing their own ideas and solutions. The Harvard Case Study Method develops this approach further by staging the information flow in a seminal case history, and asking students to draw conclusions based on what they know at each stage in the development of the case history. The essential point is that each student learns by working with the material individually, reflecting and then confirming his/her understanding. In the process students develop their knowledge and thinking skill through presentation and in discussion with an expert and with fellow students (Figure 5.3).

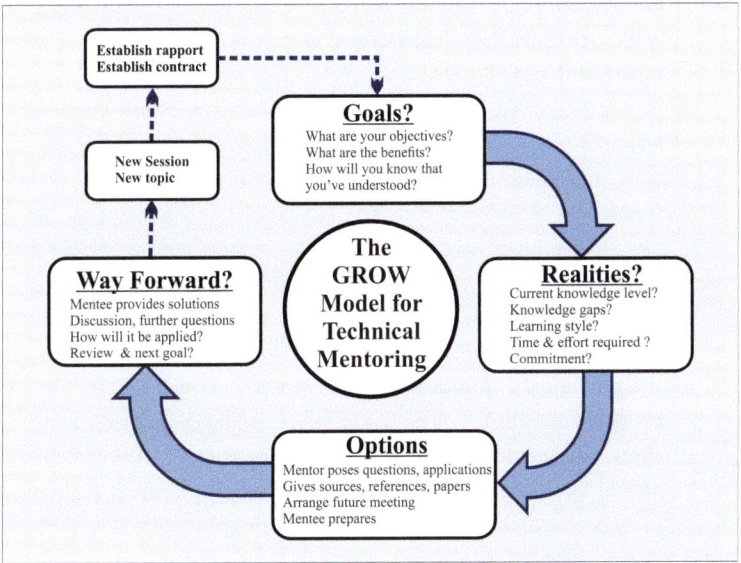

Figure 5.3 The GROW coaching model applied to technical mentoring.

EXERCISE 5.1 – USING THE GROW COACHING MODEL FOR TECHNICAL MENTORING

For technical mentoring with a mentee or a small group at the same stage of development (maximum four), try using the GROW framework for <u>two</u> tutorial-style sessions covering the same subject. (See also Section 4.2 above, and Figure 5.1).

Session one – establish the subject

1   *Building rapport* – Create a relaxed, psychologically safe atmosphere conducive to learning. Ask a few questions to get an idea of the mentees' current levels of knowledge, cognitive skill and ranges of preferred learning style (see Section 5.2).

2   *Contracting* (see Section 4.3) – Describe the method and set-up as follows to the mentees. Agree the time (30 minutes). Say what you can and can't do. Clarify mutual expectations, especially the principle of self-responsibility, assure the mentee of confidentiality and negotiate holding to account.

3 **Goal** – Aim to create clarity about what it is that the mentee wants to learn, for what reason, what will be the benefit and how will they know that learning has been achieved.

4 **Realities** – Ask questions to identify the mentees' current levels of knowledge of the subject and learning style. Avoid creating the appearance of a threatening interrogation.

5 **Options** – Pose two or three of questions or problems in the technical subject area agreed, ideally relating the material to the mentees' own work in their current role. Direct them to sources of information such as papers, articles and textbooks. Also explore the options for practical work.

6 **Way forward** – Ask each of the mentees to prepare a short informal presentation giving the answers, evidence and reasoning about the problem. Agree a time to meet again, say two weeks hence, for the presentations and a discussion. It could be in a meeting room, in the field, the laboratory or in a clinical setting as appropriate.

Session two – exploring the subject

7 **Rapport and contracting** – re-create a learning atmosphere and re-state the original agreements. Agree a time (90 minutes).

8 **Goal** – Re-state the original goal. Check if anything has changed for the mentees as a result of their private research studies.

9 **Realities** – Ask the mentees to address the questions as set, with reasoning or perhaps giving the pre-arranged presentations.

10 **Options** – Stimulate discussion and facilitate by asking questions, respond with reflective listening and give further information as needed. Use the Socratic questioning technique if appropriate (see Section 5.3). Explore practical applications in their current work.

11 **Way forward** – Finally, review and acknowledge the mentees' progress in their understanding. Ask what they want to learn next and repeat the cycle.

5.3.1 The Socratic questioning technique

The best way to learn any new skill or behaviour is to learn under an expert. This is one of the most important conclusions of Social Learning Theory (Bandura, 1977). When, as learners, we see someone exercising a skill, we remember the sequence of behaviours and try to replicate this in order to develop our own skills. We gradually develop proficiency through continual practice with the expert looking on, encouraging, correcting and advising. In this, the expert acts as a role model. Other possible ways of learning a skill, such as experimenting on our own or reading about it, can work but are relatively inefficient and may even be ineffective and de-motivating. For practical skill training, as opposed to accumulating technical knowledge, the best way to learn is by working with people who are already competent under real world conditions.

The Socratic questioning technique is a highly effective way of developing the characteristic cognitive skills of any profession. The method is named after the classical Greek philosopher Socrates, who taught in Athens in the fifth Century BC. We know exactly how Socrates proceeded because his method was recorded in a series of dramatic dialogues written by his follower Plato.

In Plato's account Socrates sits in a group, choses someone who has a different point of view to his own and asks him a philosophical question. He then conducts a questioning sequence that brings that person round to Socrates' point of view. Although Socrates' questioning is with a single individual, the other members of the group identify with the student as they listen and follow the process for themselves.

In a group of mentees, ask for a volunteer. Start with a specific technical question or problem about which you think your volunteer will have an answer or opinion. Your initial question is effectively a test for a generic level of knowledge in the hierarchy of knowledge in Bloom's Taxonomy as described in Section 5.2. Start with a level of knowledge at which you think the volunteer and the group as a whole feels secure. Then engage in a questioning sequence designed to reveal and develop knowledge progressively. It may be necessary to give some additional information at each step or revise something they already knew but had forgotten. The endpoint is when the volunteer arrives at a significant conclusion or cognitive skill that is new to him, and ideally to the whole group.

The mentor's skill lies not in following a recipe or list of questions but in responding flexibly and supportively to the volunteer's personal thinking in the moment. An apt metaphor from executive coaching is 'dancing in the moment'. The interaction takes time, patience and kindness. Be sure not to force your solution. Ideally, the volunteer should arrive at a solution for himself at which point you will see the 'lightbulb moment'. A good learning outcome is indicated by spontaneous questions from volunteers and energetic general discussion involving the whole group. *Socratic questioning models the classical scientific method by following the sequence of: hypothesis – test – observation – conclusion - repeat.* It is therefore especially useful for cognitive skill development in science and technology. Case History 5.1 illustrates the method by example.

One caveat is essential. The Socratic questioning technique will be dysfunctional and counter-productive if the volunteer perceives it as threatening or intimidating. As a general rule, leading questions should not otherwise be used in mentoring because they are covertly directive rather than non-directive (see Section 3.2). To ask leading questions in the way a barrister conducts cross-examination in a court would be counter-productive in a mentoring situation. The tone and manner of the interaction is crucial to how well the method works, especially in a group. It is absolutely essential to have previously established a basic level of trust and rapport, to be tactful, respectful, patient and understanding and to observe the basic principles of client-centred mentoring. In the Introduction to his translation of four of Plato's dialogues set around the trial and death of Socrates, Tredennick emphasises Socrates' humanity, "…his kindly heart, his quick perception, his unfailing tact and patience and cheerfulness – all enlivened by an impish sense of humour – made him an ideal companion" (Tredennick, 1959). Clearly Socrates was also an ideal mentor.

CASE HISTORY 5.1 – EARLY CAREER SCIENTISTS TURN THEORY INTO PRACTICE (JWA)

Setting: A group of five junior geologists with a mentor (JWA) in a large open pit gold mine in Ghana. The group are early career Ghanaian national graduates in training as mine geologists.

Building rapport: The mentor and group chat casually on the way to the site and get to know each other.

Establishing the contract: Arriving at a rock face, JWA asks what the mentees most want to learn. They ask about structural mapping. He already knows that rote learning is the commonest training method in Ghana, and also that both extreme deference and shyness are normal among early career Ghanaian professionals. He explains that he is going to ask a series of questions with the understanding that there's no such thing as a wrong answer and asks for agreement and co-operation. Given indications of agreement, JWA calls for a volunteer. They all know that one member of the group, Kofi (not his real name), has already done some unsupervised pit mapping and so should be in the best position to answer questions. After some hesitation and vigorous persuasion from the others, Kofi steps forward. The mentor encourages anyone to join in if they want to, knowing that they will be too shy at first.

The Socratic Questioning Method in use:

JWA: Thank you Kofi. Let us begin by asking, how would you start to make a geological map of this bench?

KOFI: I would stand back and look for structures.

JWA: OK, let's just do that. Now, please tell me what you see.

KOFI: There's this structure here and this one and this [points to a shallow-dipping fracture plane cutting across and displacing a band of rock]

JWA: Fair enough. What are these structures called?

KOFI & ALL: [shy or embarrassed silence]

JWA: Don't worry. Let's work it out from first principles. This band here is clearly different to the main body of rock on either side. What do you think it is?

KOFI: I've seen this in drill core. I think it is a bed of chlorite schist.

JWA: Fair enough. So we know that a bed is a parallel-sided layer of rock. What about this plane that cuts across the chlorite schist bed? What might that be?

KOFI: Umm … could it be a fault? [thoughtfully]

JWA: How you could tell that it is a fault?

KOFI: [Gives rambling account of basic theory of stress regimes with extension and compression vectors]

JWA: How would you apply that theory here?

KOFI & ALL: [Puzzled expressions, thinking]

JWA: [Stays silent for a minute. When it is clear nobody will answer, prompts -] What do faults do when they cut across other structures?

KOFI: [tentatively] Displace them?

JWA: Can you see any displacement here?

KOFI: [hesitation] Yes ... [points out displacement along the fault plane. Others excitedly join in]

JWA: Great. So you've correctly shown that this fracture plane is a fault. Now recall the stress regime in classic Andersonian faulting – normal, thrust and strike-slip. [Blank looks. Reminding the group of what they learned in college, JWA sketches the Andersonian scheme on a drawing pad] So, which type of fault is it?.....

KOFI: The fault has moved the chlorite schist bed from here to here. So it must be a thrust fault.

JWA: Well done! You have worked it out from first principles. Thank you very much Kofi. Now, how does that relate to the position of the orebody over there in the west end of the Pit? [Another volunteer steps forward and the interaction continues with enthusiastic group participation]

Review and Action: After about three hours in the field the group returns to the mine office. Mine maps are put on the table and discussed over a cup of coffee. JWA asks each person to say one thing they learned in the field. They have an animated discussion about how they are going to use what they've learned in the different sectors of the mine in which each person is currently working. **Note**: In this example Kofi recalls the first principles and is led to work out the solution step-by-step for himself. The Mentor's behaviour is encouraging and he takes care not to put down the volunteer when he can't answer. His colleagues learn vicariously as they look on. In this way they all learn how the basic thinking process works in practice, something that seems to have been missed in their undergraduate training.

5.4 PROFESSIONAL ATTITUDES AND MOTIVATIONAL INTERVIEWING

To follow a profession is to have a vocation, a calling, one that generates a strong feeling in the practitioner. A vocation is not a matter of casual choice, it entails a sense of dedication, of taking responsibility for one's own decisions and the path one choses to follow in life. *Your mentees' attitudes and motivation to practice their chosen profession will determine how their careers develop.*

Professionalism is a rational belief system based on a set of attitudes and values that are held in common by the members of a profession. By agreement, often over centuries, professional institutions lay down the codes of ethics and competencies that characterise their own professions. *An attitude, or mind-set, is defined as a settled way of thinking or feeling about something* (Oxford English Dictionary, 1964). Attitudes are binary and based on emotions of attraction or aversion that allow us to make judgements and decisions. Attitudes are formed through our individual personal life experience. They are derived from our individual values and beliefs, many of which were formed at an early stage of our lives and have since descended into our sub-conscious. But they are abstract ideas; we cannot see attitudes in other people. The best we can do is to observe the behaviours that we can see in other people and make inferences about the attitudes that create the behaviours.

To make progress in science, engineering and technology it seems that six attitudes are essential, (1) self-honesty, (2) enthusiasm, (3) curiosity, (4) self-motivation, (5) self-belief and (6) ethical professionalism. Table 5.3 lists examples of observable positive behaviours that indicate possession of each of the six attitudes. They form a kind of baseline of positive professional behaviour. If your mentee displays these behaviours, then obviously that behaviour should be acknowledged and encouraged. However, if he/she does *not* display the behaviours that indicate these desirable professional attitudes, or even displays the opposite behaviour, then something is wrong. His/her professional practice will inevitably suffer. As a mentor you have the responsibility to draw the mentee's attention to what you see as dysfunctional behaviour and what you think that implies. So, for example, a lack of self-honesty will be revealed by dishonest actions, for example in plagiarism or credit-taking. A lack of curiosity may be apparent in someone who has a fixed

Table 5.3 Some essential attitudes and associated behaviours in
science and engineering

Attitudes and values	Examples of observable positive behaviours
Self-honesty Truth, independence	• Respects the evidence and necessary conclusions • Differentiates observations from interpretations • Willing to acknowledge his/her own error • Takes all relevant considerations into account impartially
Enthusiasm Energy, joy	• Exhibits energy and determination in investigations • Readily engages in lively discussion • Self-expression demonstrates love of the subject
Curiosity Knowledge, creativity	• Open to new ideas • Enquiring manner in conversation • Identifies technical problems as challenges • Looks for opportunities to gather new information • Enjoys creative thinking
Self-motivation Autonomy, self-development	• Takes action (within limits of delegated authority) • Willing to look actively for solutions • Needs little direction beyond initial briefing • Strong focus on the issues at hand, not easily distracted • Takes responsibility for his/her own professional development (CPD)
Self-belief Moral courage, self-determination	• Displays appropriate confidence in communicating technical ideas • Contributes positively in debate • Able to challenge colleagues' ideas respectfully and defend his/her own
Ethical **Professionalism** Ethical belief, communal responsibility, fairness	• Uses best efforts in all circumstances • Behaves honestly in all circumstances • Puts clients' and stakeholders' interests ahead of his/her own • Acknowledges contributions of fellow scientists and collaborators • Collaborative; helps colleagues with their own professional development • Keeps privileged information confidential • Treats all fairly and without prejudice

opinion or does not want to develop his/her knowledge. Prejudice against people of another race, religion or gender would indicate a lack of fairness that otherwise should be present and characteristic of ethical professionalism and also of self-honesty. If your mentee shows a lack of positive behaviour or even dysfunctional behaviour, then use the routine for structured feedback sensitively and tactfully (Section 3.4). An emotional response, possibly quite negative, is likely and so careful self-management on your part is essential. *In general, as a mentor you will be acting as a role model so the behaviour that indicates your own attitudes is paramount.*

CASE HISTORY 5.2 – FOLLOW YOUR PASSION (TIMOTHY BRUNDLE)

Sir Richard Barnett was the Vice Chancellor of Ulster University until 2015. He was the first member of his family to go to university and believed that every young person should have opportunities irrespective of their family background. He recruited me as the Ulster's first Director of Innovation to help put the University's knowledge to work on behalf of the economy and society. His advice to students and staff was to 'follow your passion and do so with determination. Ignore the naysayers. All successful people work bloody hard.'

Sir Richard felt the idea of innovation was preposterous. We could agree that technology allows us to do things in ways that weren't previously possible, but he would cock his eyebrow and remark that this principle has been true since humans first created fire. For over a thousand years we within universities had found that small research groups, that are multidisciplinary and maintain diversity, that are collaborative and international in their outlook, are those which have resulted in the technology breakthroughs that make society a fairer, healthier and better place. These principles excited him, as did the way in which the internet had made the tools available for experimentation more evenly distributed. Richard taught me that creativity and humanity have always been the source of our enlightenment and we must take care that we don't design out the things that make us special.

EXERCISE 5.2 – PROFESSIONAL ATTITUDES

Work alone or with a partner and take turns. Start by reflecting on the attitudes that underlie your own professional work and how that shows in practice in your behaviours, communication and work style. Consider the behaviours &/or actions you have observed in a younger colleague or a mentee (use fictitious names). Refer to Table 5.3 for examples. Add some more attitudes and values to those listed below as you wish.

- Self-honesty (Value = truth, independence)
- Enthusiasm (Value = energy, joy)
- Curiosity (Value = knowledge, creativity)
- Self-motivation (Value = autonomy, self-development)
- Self-motivation (Value = autonomy, self-development)
- Self-belief (Value = moral courage, self-determination)
- Ethical Professionalism (Value = ethics, responsibility)

Write brief notes comparing your behaviours that express those attitudes with those of the colleague or mentee. Use keywords to indicate the specific context and give examples of the behaviours or actions

Caveat – Mental illness is strongly associated with dysfunctional behaviours, although the reverse is not necessarily so. Mental illness is surprisingly common but not often recognised in the workplace. An estimated 1 in 6 of the UK population has suffered a mental health disorder. The commonest are anxiety and depression with 7.8% of the UK population meeting the criteria for clinical diagnosis (Mental Health Foundation, 2020). In fact, even if the great majority of us do not require clinical treatment, just about everybody suffers from an emotional disturbance at some time in their lives. In a mentoring context compassion about mental illness is an absolute necessity. Empathic listening skills are at a premium in these situations. The very worst things you can say are, "Pull yourself together!", "Button up and get on with the job!" and, "If I was you I would …". If one of your mentees shows clear signs of mental distress, then you have a duty of care to recommend that they seek help from a qualified counsellor or therapist. Any attempt at treatment by an unqualified person, however well-intentioned, risks exacerbating the problem. In this situation you should refer back to your agreement in scoping and contracting about referring on (see Section 4.2), then carefully follow the structured feedback procedure (see Section 3.6) and say something like, "This situation is beyond my professional competence. In your best interest, I think you should speak to X." What follows assumes that mentees are *not* suffering mental ill-health.

5.4.1 Motivational interviewing and the cycle of change

Although self-motivation is an essential aspect of professionalism, we have all met people who seem to be blocking their own progress. If you find yourself faced with a mentee whose behaviour indicates that he/she lacks motivation, what should you do?

Motivation is the ability and energy to sustain a positive atti-tude. If a mentee comes to you with say, a technical subject and seems to be de-motivated you should set aside the subject of technical mentoring and address the motivational issue. Without that, little else can be achieved. *It is important to understand that we cannot change another person. People can only change them-selves. So be aware that you cannot 'fix' another person's attitude.* A good professional relationship can exert a strong influence but that is as far as it can go. As a mentor the best you can do is to hold a rational, adult-to-adult feedback conversation. However, persuasion is unlikely to solve an attitudinal problem by itself. You might be able to facilitate a change in your mentee by using the motivational interviewing technique but success cannot be guaranteed.

The motivational interviewing technique (MI) was developed by Miller and Rollnick in the 1980s for addiction counselling and later adapted for coaching psychology by Passmore and Whybrow (2007). In following the Self- motivation Principle, MI relies on developing intrinsic motivation rather than using classical behavioural rein-forcement based on extrinsic motivators such as salary, status and acknowledgement. MI also recognises that no change of attitude or any major decision is instantaneous. Change is one part of a cy-clical process that can be described in four stages (see Figure 5.4).

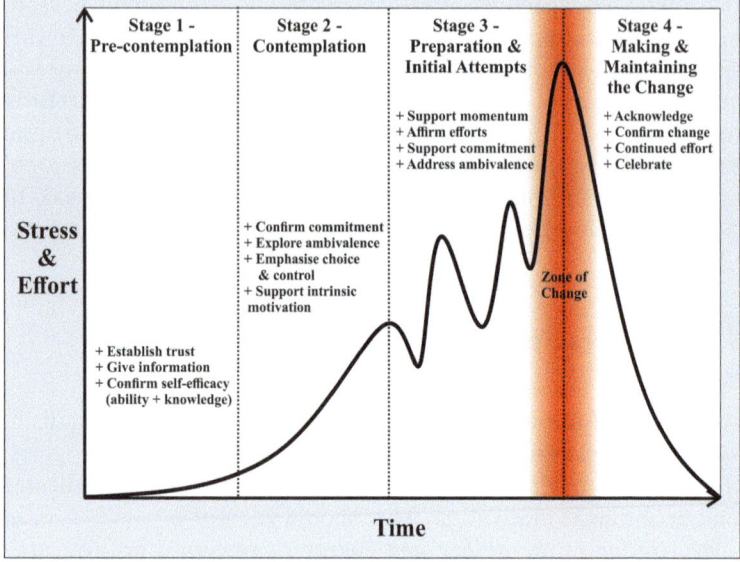

Figure 5.4 The four stages of the cycle of change.

The art in using MI lies in understanding where your mentee is on the cycle of change and then adapting your style and approach to your mentee's efforts and their changing level of commitment accordingly.

Before starting motivational interviewing there are some essential preconditions:

- Establishing and maintaining trust are essential conditions for motivational interviewing. The chances of success are small unless you have already built up some level of trust.
- Effective self-management, including strictly maintaining your own professional ethics and boundaries, is essential. Before beginning get clear with yourself about your own intentions.
- Communicating appropriately includes taking care not to use judgemental words or non-verbal communication that could elicit a defensive response. MI has its origins in client-centred counselling and *its most important technique is reflective listening* (see Sections 4.1 and 3.3).

The four stages of the cycle of change are:

1 *Pre-contemplation* – At the start your mentee is unaware of the need for a decision or a change. In working with him/her previously you will have gathered some factual evidence. You have listened not only to what he/she has said but also heard the tone of voice, energy and word choice. You have seen the outcomes of various actions and behaviours. From that you have inferred the attitude and level of motivation that *might* be causing the behaviour and the outcomes you are seeing. Ask your mentee for permission to talk about his/her attitudes, values and level of motivation. This will lead naturally to a structured feedback conversation which will let him/her know your opinion about the change that they need to make. In this it is essential to confirm his/her *self-efficacy* – the belief they have in their personal ability to do whatever is required to make the change happen. If they don't have that essential self-efficacy, then change cannot happen. If they have at least some self-efficacy, then present the necessary information and argument, making sure your own behaviour is compassionate and non-judgemental. Success depends not only on rational argument but also on the degree of trust placed in you. In following the cycle of change it is important that they give you permission to hold them to account.

2 *Contemplation* – In the next stage the mentee weighs up the costs and benefits of change and examines the arguments. *Ambivalence and resistance are inevitable*. Your patience and tact in dealing with his/her frustration and difficulties are essential means of support. Emphasise his/her self-efficacy, personal choice and control. Clarify the rational arguments in a non-judgemental way. Use reflective listening techniques.

3 *Preparation & initial attempts* – If your mentee carries on to the next stage he/she will make a first tentative effort to change and may fail. Help by exploring all possible sources of intrinsic motivation. Although extrinsic reward can help, developing intrinsic motivation will always be more powerful. What really interests your mentee? What do they risk losing by making the change? What's really at stake? How desirable are the benefits? What will success look like? What strong values do they hold? Do they care about what family or colleagues think? How can these be connected to drive change? You can enhance

intrinsic motivation by asking your mentee to recall one or more *reference experiences* in vivid detail – that is, previous experiences of overcoming obstacles successfully. Repeated recall of reference experiences with support and affirmation are essential after each failure. Encourage determination. Recall the objectives and benefits and reaffirm commitment. You will probably need to go back over the actions in stage 2, perhaps on several occasions and build up to another attempt.

4   *Making & maintaining the change* – Finally, the mentee appears to have succeeded in making the change but be aware that it is not yet securely embedded. Continue with support and checking. Acknowledgement and closure are essential. When success is finally assured, and not before, celebrate! Closure is necessary in order for the change to be securely embedded.

CASE HISTORY 5.3 – FEAR AND ANXIETY INHIBIT A MENTEE'S CAREER (JWA)

Achieving Chartered status is widely considered to be an onerous spare-time task. It includes writing an application, preparing a professional report, putting together a package of supporting documents and finding sponsors, all over a period of at least three months and quite often much more. Some years ago an applied geoscientist, who we'll call Stanley (not his real name), came to me for help. In spite of fulfilling all the entry qualifications and plainly being a worthy candidate, Stanley had great difficulty in getting on with his application by himself. He was a very decent, conscientious, if somewhat over-anxious character who normally gave 110% to everything he did. I helped him in making detailed plans, schedules and targets but whatever we tried, he simply couldn't stick to it. He always had a good excuse for not having stuck to his plan – new projects and responsibilities at work and needing to spend at home with his wife and their new baby. Eventually he seemed to give up on the idea of Chartership and we reluctantly parted company. I was stumped and felt that I had failed Stanley.

The back story is that about ten years before we met, Stanley had had a career as a laboratory technician. For many years his hobby of fossil and mineral collecting had been his real passion. As an amateur he began to study geology seriously. At some point he decided to try for a change of career and become a professional geologist. Making a huge personal commitment over five years as a part-time student, Stanley took a degree in earth sciences in a local university near his home. On graduating he took an even bigger risk, left his technician's job and took a less well-paid job working for a large engineering company as a junior geologist on a short-term contact. Stanley was highly motivated and, being older than others at the same level, he was more mature in his attitude to his work. He performed well, was taken on as a permanent employee and promoted to a more senior role. Stanley had successfully navigated a major career change. He had enhanced his education, got a good job, achieved a measure of security for his family and seemed to have a good future ahead.

Some years later, and quite coincidentally, Stanley's employers declared a policy change that was to affect him crucially. The directors had become increasingly concerned with an industry-wide threat of claims of liability for professional negligence. In an attempt to defend their company, its directors declared that all professional employees at or above a certain grade must apply for Chartered status. That just happened to be the level to which Stanley had been recently promoted. To begin with, Stanley's line managers were not very diligent in enforcing the new policy and so he passively avoided applying for Chartership. However, that left him feeling insecure in his job and so he was working all hours to try and prove himself. Meanwhile, the vague threat of job insecurity was left unresolved. Then he and his wife had their first baby. As time went on Stanley became more exhausted by the internal conflict between his job, his duty to his family and his own professional development. He couldn't think clearly and reacted impulsively to whatever task was in front of him. He suffered fear, a lack of self-belief, guilt and anxiety that brought on depression and with it, de-motivation. With every day the Chartership mountain that he could have easily climbed at the start just got bigger and bigger.

By chance a year or so after we had tried to work together, I met Stanley at a lecture meeting. I was delighted when he told me that he had sought professional counselling that had successfully helped him to get through depression. He had recently completed the application process and had been awarded Chartership, a well-deserved but very hard-earned qualification. In hindsight I realise I could have served him better if I had obtained his permission to use a motivational interviewing style, something I didn't know much about in those days. Before making plans for his Chartership application we would have needed to focus on the ways he was holding himself back and how the process of change could happen for him.

5.5 MENTORING FOR PROFESSIONAL QUALIFICATIONS AND ACCREDITATION

Nowadays a primary degree in science, technology or engineering is no more than an entry qualification to a junior level of practice in a profession. For any career to progress to higher levels the practitioner must acquire some form of accreditation awarded by one of the learned societies or professional associations. Institutions and learned societies in the UK that have been incorporated under a Royal Charter offer accreditation in form of Chartership (*e.g.* CEng, CSci, CGeol)., but there are also many other varieties of accreditation. In whatever form it is awarded, accreditation is a recognition by peers that the holder has established a level of competence in the profession. Accreditation reassures employers, clients and colleagues that the holder has the approval and support of peers, and that he/she has kept up-to-date with recent advances in their field. An increasing number of science and engineering organisations now insist that their professional employees are accredited. In so doing they usually also offer mentoring schemes to help more junior staff gain the necessary qualifications. How should mentors tackle what has become one of the commonest mentoring engagements in the STEM fields?

Mentoring for professional qualifications can be tricky because it ***requires both directive and non-directive styles***. The art lies in finding the right balance. On one hand accreditation mentoring is directive in that you have a responsibility to help your mentee achieve the required level of technical ability and to point out anything obviously wrong with the application. On the other hand, mentoring must be non-directive in that you must not tell the candidate exactly how to fill out the application and you cannot control the progress of an interview or examination. *The mentor's intention should be to help the candidate develop a level of self-awareness and communication skill that leads them to present their professional competencies in the best light.*

Accreditation mentoring begins in one of two ways. One is that an applicant comes to you for help because your name is on a list of mentors. The other is that an applicant is matched with you as a mentor within a training scheme. In either case both you and your mentee should **understand and accept the principle of self-responsibility** (Section 4.2). Although mentoring can be a huge help, the candidate's has to get through by their own efforts. You are not responsible for 'getting them through'. *As a mentor your job is limited to facilitation.*

Although the STEM fields encompass a huge variety of professional qualifications, most professional institutions have very similar arrangements for gaining accreditation. There are commonly four stages in the accreditation process: (1) Preparation, (2) Application, (3) Validation and (4) Decision.

Mentoring presents different challenges at each of the four stages of the process. In fact, the accreditation process can be mapped directly onto the Cycle of Change. Similar types of mentoring support are required in each stage. A flow chart showing steps in the process is shown in Figure 5.5.

5.5.1 Stage 1 – Preparation

There are always some pre-qualification conditions for accreditation. The applicant must already hold some academic qualifications, usually tightly specified. He/she must have been in professional practice for a specified duration, will have had certain types of experience, can provide a record of continuous professional development and can call on sponsors who are already accredited to certify that he/she is a suitable candidate. This stage is equivalent to the pre-contemplation stage of the Cycle of Change (see Figure 5.4).

As an accreditation mentor obviously you must yourself hold the accreditation and be familiar with the learned society's regulations. As in any other mentoring engagement, your job begins with connecting, scoping and contracting. Many young professionals who are eligible for accreditation have never before considered it and so the first step is to tell them about professional status and its

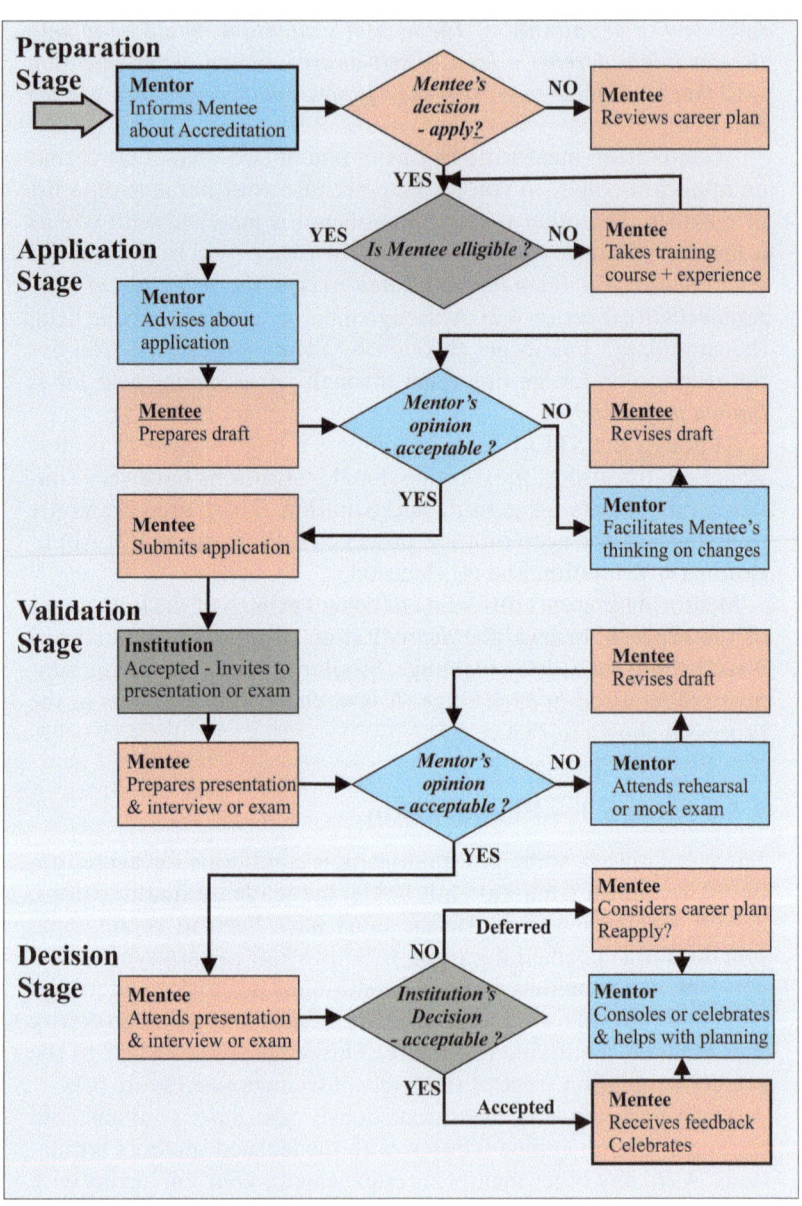

Figure 5.5 Flow chart for accreditation mentoring.

benefits. If the mentee is part of a company training scheme, then mentoring pairs will be set up for you in the matching process. If you are meeting a candidate for the first time, then treat it as a scoping meeting and begin with discussion about mentoring procedures and principles, and especially self-responsibility (see Section 4.2). Also point out that neither of you is obliged to engage in mentoring with the other. Similarly, if your potential mentee decides that he/she does not want to pursue accreditation, then that is entirely their personal responsibility. In this case, and depending on the employer's need for accredited staff, you should encourage the young professional to consider the future direction of their career.

If you have agreed to work together, then the first thing is to *establish whether or not your mentee is eligible for accreditation*, usually a straightforward matter of verifying that the mentee holds appropriate academic qualifications, has completed a defined period of post-graduate experience of professional practice and is able to demonstrate a specified set of competencies. Only two outcomes are possible. *On the one hand, if he/she is <u>not</u> currently eligible*, then it normally means designing a programme of supplementary training for more appropriate qualifications and working on specific jobs to get the necessary experience and reports. That will probably need help from others, including trainers and managers and could involve a long term mentoring engagement for you both while the mentee works through the training and experience. How much time and effort is actually required from you as a mentor will depend on what agreements you make during scoping and contracting.

On the other hand, if your mentee is eligible for accreditation, then he/she will need to make the commitment to preparing and submitting an application. For that a feasible plan for arranging sponsors, gathering the evidence for competency, obtaining permission to use the supporting documents, writing and revising the application and planning the interview will be required:

- Confirm that the mentee wishes to make a firm commitment to gaining accreditation
- Ask mentee to prepare a detailed plan including a schedule and time budget; 3 months is common duration
- Go through the plan together and explain any changes needed. Candidates often underestimate the amount of work they will

need to put in, especially in assembling documents, writing the required report or essay and preparing for an examination or interview

- Negotiate the number and spacing of mentoring meetings within the preparation schedule, say, four 90 minutes sessions spread over three months
- Check arrangements for sponsors and, if necessary, advise the mentee to ask the institution for advice and help. Check that they have notified the institution of their intention to apply.
- Contracting will include an explicit permission for you to hold your mentee to account for progress. Your role is to check they are on target. If not, then to understand what is causing the delay and help reschedule.
- Ask the mentee for an explicit and clear commitment to these arrangements.

5.5.2 Stage 2 – Application

Stage 2 begins with the mentee's actions including finding sponsors, filling out the application form, completing a professional report, collating a package of professional documents, paying a fee and notifying the sponsors that the application is in progress. The mentee gathers evidence and carefully constructs the arguments for competency in written form. As mentor your job is to facilitate the mentee's thinking about how to make the arguments for competency, but *not* to do it for them. The institution will have arrangements to check that the application really is all the candidate's own work. An argument for mastery of a set of competencies is not a simple matter of presenting a CV. It is a logical statement of mastery together with accounts of selected events and pieces of work that support the claim to the candidate's mastery. If necessary, help your mentee to understand that a well-formed argument must show the attributes of clarity, logical validity, consistency and precision. Accrediting institutions complain that a lack of checks on spelling, grammar, references and acknowledgements reflects poorly on the candidate's written communication competency skills. As their mentor you will need to be able to allocate time for close reading of the draft documents. You will usually have to recommend revisions, sometimes repeated more than once, while emphasising that the decision on the content of the application and when to submit must be the mentee's own responsibility. Mentoring sessions will normally conclude with agreements about the next steps and homework to complete each of the steps.

Finding time and energy to prepare an application for accreditation can be a serious challenge for any candidate. If your mentee gets bogged down he/she may procrastinate. Don't ignore their problems or brush them off. Acknowledge the difficulties and help the mentee work out strategies to address the problems. Find the proper balance of motivational, advisory and listening skills appropriate to the second phase of the Cycle of Change. Your mentee might say, "... I can't possibly find the time"; "What happens if I fail?" *Ambivalence and resistance to the accreditation process is normal and should be expected during this stage.* Whether long or short, the process will follow the Cycle of Change and so you may find motivational interviewing techniques useful. For this work to be effective it will be essential that you have already obtained the mentee's permission to hold them to account. Without first getting their permission, any attempt at motivation will fail.

5.5.3 Stage 3 – Validation

The professional institution will have some process for validating the application. The primary purposes are to test the arguments made in the written application, to check that claims made for competency are supported by sound evidence and that achievements have not been exaggerated. As the candidate makes his/her case for competency, the panel will also be looking for evidence of independent thinking, clarity of expression and other communication skills. This stage includes investigations such as contacting sponsors and checking professional documents and reports submitted in evidence. Validation often includes a presentation by the candidate to a panel of colleagues in the same field, as well as an oral examination (*viva voce*) and, in some institutions, a written examination and /or essay.

Mentoring should provide a safe learning environment for understanding and developing the full range of professional communication skills. Effective communication is a consequence of clear, rational thinking. Providing constructive advice and supportive feedback is also necessary in this stage, but can be even more sensitive when it is about self-presentation and self-expression. Support the mentee in identifying questions that they find hard to answer, help plan what to say if the hard questions come up and how to deal with the anticipatory anxiety. The mentee will almost inevitably suffer performance anxiety and indeed, a low level of anxiety is a helpful stimulus. You can set up a role-play interview and presentation with you taking the role of a panel member. Give feedback on the presentation and ask questions you think are likely. Video recording provides a very effective form of feedback when your mentee can see and assess his/her performance for themselves. If necessary, advise the mentee to attend a presentation skills training course.

5.5.4 Stage 4 – The decision

Finally, the institution makes its decision on the application. For the candidate, Stage 4 is the final stage of the Cycle of Change. After the panel have completed their investigations in the validation stage, control of events passes from the mentee to the professional institution as they consider and then formally notify the mentee of their decision. For most institutions there are only two outcomes, either acceptance or deferral (usually not expressed as refusal or failure). Additionally, they may offer written advice about how the mentee could make improvements. Whatever the outcome, as a mentor you have a responsibility to support your mentee in dealing with the consequences.

If you have scoped and contracted the mentoring engagement correctly right back at the start, then your mentee will take personal responsibility for the outcome. In the event of deferral, your role as mentor will be essential in providing your mentee with support and encouragement not to lose heart. He/she may or may not want to continue working with you and to re-apply. In the event of acceptance, then acknowledgement and closure are psychologically necessary. Help your mentee to celebrate. Note that the principle of self-responsibility still applies. That means *you cannot take any credit for the mentee's success*. In either case, encourage your mentee to review his/her career plan.

5.6 MENTORING FOR SELF-CONFIDENCE

The industrialist Henry Ford is alleged to have quipped, "Whether you think you can or whether think you can't, you're right." As a mentor, you will encounter mentees who, despite being intelligent and well-motivated, experience an unexpectedly high degree of difficulty in some area of their professional practice. Presenting before a live audience is one of the commonest problems of this kind. The issue will be expressed with a sense of tension, fear and heavy emotional loading. If you look closely you'll usually discover that a lack of self-confidence is the underlying problem. Lack of self-confidence and low self-esteem are surprisingly common and usually unacknowledged. They are potentially very serious and can be the cause of a very wide range of problems at work including stress-related ill-health, absenteeism, employee disengagement, executive burn-out, interpersonal conflict, poor supervision, inadequate performance and many other problems besides. For many people, a lack of self-confidence simply blocks professional development. Conversely, the sense of achievement in getting over the block can be an enormous help in developing an individual's professional ability and subsequent progress. As a mentor, how should you help a mentee who suffers from a lack of self-confidence?

Caveat – Severe anxiety and depression can be associated with a lack of self-confidence and may require clinical intervention. As a mentor, you have responsibility to refer on a chronic or acute sufferer to specialised health services. In severe cases it is unlikely that you can make a real difference without specialised knowledge and counselling skill and you risk making things worse. The best you can do is offer carefully and sensitively delivered feedback, find out how to access professional services and recommend that the victim seeks that help. Again, the 'referring on' procedure applies (see also Section 5.4). The rest of this section assumes that your mentee enjoys a normal level of mental health.

Apart from this extreme situation, a mild to moderate lack of self-confidence in specific circumstances is very common and can be successfully addressed in mentoring. The confidential and supportive circumstance of mentoring is a great place for your mentee to create and explore their awareness of their own problems around self-confidence.

5.6.1 Understanding what self-confidence really means

The Intrinsic Motivation Principle asserts that we are all intrinsically self-motivated (see Section 4.1). Our natural state should be possession of a stable and moderately high level of self-confidence and self-belief. In spite of this, everyone experiences a loss of confidence at some time in our lives. Lack of self-confidence and low self-esteem may have their foundation in adverse experiences during childhood and can also be initiated in adulthood. We may have been severely criticised in a meeting, failed a vital test, lost an important contract, said something that we later regret or rejected for a job we expected to be given. In some way or another, our self-confidence has taken a knock. We may feel that we have lost status and respect. Our inner critic yells at our inner child. As we replay old fears in our mind's eye, the negative load is added to as the flashbacks increase in intensity. Subsequent repetitions of similar negative events can create a self-fulfilling downward spiral in a chronic lack of self-confidence.

Self-efficacy refers to the belief that we can perform a specific task. The inventor of Social Learning Theory, Albert Bandura, said, "Self-efficacy is concerned, not so much with the skills one possesses, but with judgements of what one can do with the skills one possesses" (Bandura, 1977). In a series of experiments Bandura measured self-efficacy and showed it to be an accurate predictor of success or failure in any task. The important point is that although lack of self-efficacy may occur in some particular area of our lives, nobody is without confidence in all areas of their lives. In fact, confidence is a learned skill and we are all confident in some areas of our lives. If your mentee has learned to be confident in one area, then he/she can learn to transfer the skill to another area.

5.6.2 Helping your mentee to build confidence

As with motivational interviewing, building self-confidence is also a matter of helping mentees to change their attitudes. Again, to overcome a lack of self-confidence your mentee must follow the Cycle of Change (see Figure 5.4). He/she has to want to make the change to more confident behaviour. As mentor, you can expect to begin by creating awareness of what the lack of self-confidence really means for your mentee. In the next stage you can move on to help contemplate what would have to happen to make the change from the current situation and then to support them in making the change. In this you can expect to support them through negative feelings and resistance. Getting over the crest of the curve and making the change to greater confidence. The whole change process is likely to take a series of sessions for the necessary awareness and understanding to emerge and for the required confidence to build.

Critical steps and questions to help build confidence are as follows:

- *Contracting* is again the most important first step. Accept that you cannot "fix" your mentee. Only they can make the change to more confident behaviour for themselves. Self-responsibility is a critical part of understanding on the path of developing confidence. You must have gained your mentee's trust before they can hear this message from you.
- *Clarify the confidence issue* – Help your mentee to gain clarity. In which particular area does he/she lack confidence? How and in what way? What exactly is his/her judgement about their own self-efficacy? What are the details of the back story, or perhaps the triggering event? What are the consequences of the mentee's lack of self-confidence?
- *Create an objective assessment* – On what evidence is their assessment of their own abilities based? Question the mentee's subjective assessments and challenge 'catastrophizing' language. Find and make appropriate comparisons with other people and their performance – there will always be some people who are better and some worse. Set a task to gather objective evidence, for example by considering achievements gained and unsolicited compliments by others.
- *Re-create a reference experience of confidence* – Confidence is an attitude. As such, it is an abstraction that is only made evident by specific forms of behaviour. A reference experience is an experience that we can recall and then copy. Ask the mentee to think back to some area in which he/she performed confidently in the past or performs confidently now. It doesn't have to be in work. Leisure activities and sports are especially rich in reference experiences. Look for real examples and ask the mentee to describe at least one experience in vivid detail. What *exactly* does confidence look and feel like? Keep asking questions about the exact details. How did the mentee develop that confident performance? Discuss the process and explore ways in which it can be transferred.
- *Visualise the goal* – What would it be like for the mentee to behave confidently in the area in which he/she lacks confidence? What would they see, hear, feel and do? Ask your mentee to visualise and describe in as much vivid and specific detail as they can. Help them to make it come alive.

- *Review and plan* – Explore whether or not the goal of confident behaviour in some circumstance is realistically achievable and is really what the mentee needs. If it seems unfeasibly far away, can you identify some shorter intermediate steps? How does confidence fit into their view of their own career development? How does it fit into their value system? Imagine what practice or other actions would be needed to take him/her from where they are now to their visualised goal. As far as possible identify indicators that mark progress.

- *Agree private study* – Set tasks that progressively challenge your mentee to behave as if he/she were confident. Counsellors say that it is easier to behave yourself into a new way of thinking than to think yourself into a new way of behaving. With practice the feeling of confidence gradually becomes embedded and habitual. Ask your mentee to identify people (real or fictional) who exhibit the kind of confident behaviour he/she lacks and to study them. Encourage your mentee to keep an evidence log consisting of daily practice, observations and reflections. It should be private to him/her alone. Its function is to give the mentee encouragement by recording objective observations of progress.

5.7 CAREER TRANSITION MENTORING

Mentees bring all kinds of personal and professional problems to mentors whom they have learned to trust, and especially when life-changing decisions arise to confront them. Among the commonest of these decisions are those involving major job or career transitions. Your mentee may be facing redundancy and job loss or might have been offered a promotion that involves perhaps an unwanted transfer. What should you do if your mentee comes to you for advice about a career transition?

First and foremost, do *not* be tempted to respond by actually giving the advice they're asking for. The least helpful thing you can do is to say, "If I were you, I would..." If your mentee appears very anxious, you might easily feel like coming to their 'rescue' – please don't! The attempt to 'rescue' someone in emotional distress is often motivated by a desire to soothe our own sympathetic feelings. Rescuing someone often leads to strained relationships if things go wrong later on. If your mentee does not 'own' the decision because you've made it for them, they may not approach the task wholeheartedly. They could harbour the belief that you forced the decision on them and end up blaming you. If your mentee is a colleague in the same organisation then be aware of possible conflicts of interest. You must be absolutely sure that your assistance is given purely in his/her best interest, and not that of yourself or your company. The whole point is that your mentee is not a replica of you and never could be. Even though there may be superficial similarities, their personality, belief system, personal experience and a great deal else is unique.

For a stable and well-conditioned outcome, it is essential that your mentee feels in control of their part in the career transition process and owns their own decisions. The most useful service you can provide is to reaffirm the principle of self-responsibility and to offer to facilitate your mentee's thinking. Avoid getting trapped into problem thinking in which you list all the problems, become more and more miserable as you examine each one and end up by concluding that the problem is insoluble. An outcome orientation is an essential mind-set in problem-solving. Employ the basic mentoring skills – skilful questioning, active listening, building trust, self-management and giving constructive feedback – and keep the mentee focussed on finding the optimum solution for this unique situation.

Any career transition will follow the Cycle of Change described above (see Figure 5.4). The steps are much the same as those used in motivational interviewing. At the beginning your mentee may be unaware of some critical information. You should ask questions that will help them to think through what information they need in order to make a decision and to go about collecting it. In later stages ambivalence, frustration, confusion and internal argument are normal and to be expected. The decision period can be emotionally turbulent if it involves uncertainty and negotiations with other people.

Honest and accurate self-assessment of one's skills, aptitude and experience is the single most important thing anyone can do when faced with a decision about career transition. That is far more likely to result in a successful outcome than researching the job market. After contracting, then questioning and reflection are most likely to be useful in enhancing your mentee's self-awareness. Any future occupation your mentee chooses must somehow honour his/her values and aptitudes. Explore which personal circumstances are relevant, including partners and dependents, minimum income needs, location and routine. Timing and resource constraints are likely to be significant. Explore which circumstances are within the mentee's control and which are outside. Suggest other people that your mentee could talk to who have appropriately specific knowledge and experience.

Arguably the most influential approach to career and vocation management is based on the **Theory of Career Choice**, progressively developed by the American psychologist, John Holland between the 1950s and 1990s (Holland, 2009; Nauta, 2010). Holland proposed that **vocational interest is an expression of one's personality**, thought of as a unique combination of six different personality types. Knowing where your mentee fits into the so-called RIASEC model (from the acronym) will help him/her to make an informed choice of occupations within a profession. The RIASEC vocational personality types are summarised as:

- **Realistic (R)** – using manual dexterity and physical skills to make and fix things, working with tools and instruments, often outdoors; typically engineers, technologists, soldiers.
- **Investigative (I)** – using investigative skills to explore, experiment, discover and solve problems; typically scientists, doctors.
- **Artistic (A)** – using words, images, or music in self-expression, creating and designing things; typically writers, artists, actors, musicians.
- **Social (S)** – using interpersonal skills to teach, train inform and serve. Also, being concerned for other peoples' welfare; typically teachers and health care practitioners.
- **Enterprising (E)** – using interpersonal skills to lead, influence, persuade and encourage. Also, to create a social network; typically business owners, politicians, managers, promoters.
- **Conventional (C)** – using systems to organise, to work accurately with numbers and data, to create and following procedures and plans; typically administrators, librarians, accountants.

The Holland Code for a vocational personality type consists of the initial letters of the first three personality types listed in order of strength to create one of 720 possible permutations. The letters in brackets in the list above represent each type. So, for example, IEC, would be the Holland Code for someone whose strongest aptitude is investigative (I). This person is also enterprising (E), well-organised and good with numbers (C). We can imagine someone who chose to study earth science because he found its investigative aspect appealing. Within that, he specialised in geophysics because computational-based science appealed to his conventional side. His enterprising character later prompted him to build a successful company carrying out geophysical surveys for ground investigations and environmental impact.

We can make a rough guess about what our mentees' top three vocational personality types are by asking them why they chose the career path, the specialisations and hobbies they are following now. For a more reliable assessment, a number of open-source tests can be found on the internet. For example, the Holland Code has been adopted by the US Department of Labor and a test, the O*NET Interest Profiler, is currently available free of charge on their website (O*NET Interest Profiler, 2021).

5.7.1 Using the GROW model for career transition mentoring

Career transition differs from many other mentoring situations because the impetus for change usually comes from an external event, such as redundancy or an offer of promotion; new circumstances to which the mentee must adapt. As in many other mentoring topics, the GROW coaching model can be very effectively used for career transition questions. Assuming you have already held a scoping meeting, agreed to the mentoring engagement and established a contract, then you can use the GROW model along these lines:

Goals – Self-motivated people need to take charge of their own careers and create opportunities. Start by helping your mentee to explore the career path he/she would like to follow over the next, say, five years. Then, investigate how that path could be created from where the mentee is now. It won't help to think only about what he/she does not want to happen, instead help them to move towards what they do want. Create outer brackets to work within. What would an ideal situation look like? What would be completely unacceptable circumstances? Make sure the goal is what the mentee actually wants and not what he/she has been told he/she should want. Then, when a goal has been created, ask the mentee to write it down in a concise, clear and positive statement. You may have to return to this step later.

Realities – The aim of this step is to raise awareness. Support him/her in looking at the hard facts of the current situation – is it a *fait accompli* or is there room for doubt or manoeuvre? If the latter, how much and in what way? Find out which circumstances are within the mentee's control and which are outside. Explore vocational interests, aptitudes, skills, values, attitudes, personal circumstances and resource constraints as described previously. When you've done all of that you might have to go back and ask them to re-state the goal.

Options – Next help the mentee to generate as many options as possible. If all of the obvious alternatives are not acceptable, can anything be changed to make one option more acceptable than others? Can you reframe the situation to allow a different view? Does it help his/her thinking to rate each alternative options on a scale on which 1 is totally unacceptable and 10 is ideal? If so, what would change the ratings for each option up or down? Feelings, intuition and memories of past experiences all colour the pending decision and influence rational thinking. Suggest other people with appropriate knowledge and experience to whom your mentee could talk.

Way forward? Difficult decisions always involve risk and uncertainty; that's why they are difficult. Risk is the product of the probability of an event multiplied by the consequences if the event occurs. Look at both terms in the equation – is there anything that might mitigate the risk? Beware of 'analysis paralysis' – endlessly investigating and continually re-cycling thoughts and getting nowhere. All difficult decisions have to be made under conditions of continuing uncertainty. At some point in your investigations you will eventually notice diminishing returns so that a little more information will make very little material difference. If a decision has tentatively been reached, ask what more information could realistically be gathered and what difference would that make? Offer support, positive feedback and encouragement to make a clear decision, but whatever it is, ensure that the mentee owns the decision. As far as possible make yourself available to give support whatever is the outcome.

With a major career decision it will not normally be possible to do all of this in one session. Indeed, usually it would be better not. Your mentee needs time to reflect and digest new information. In fact, most of the real work will happen in the mentee's own investigations and reflections outside the mentoring sessions. A series of three or four 60-minute sessions with intensive private study and reflection in between is likely to give the best outcome.

5.8 INTERCULTURAL MENTORING

Science, engineering and technology are international. Sometimes, maybe frequently, we will receive overseas visitors in our offices, laboratories, clinics, engineering sites and educational institutions. Equally, many of us will have spent time working overseas. As part of a knowledge transfer agreement you may have been asked to mentor a colleague who comes from a different culture than your own. Assuming that language itself is not a barrier, how should you manage the communication difficulties that can arise due to cultural differences?

An apt metaphor about the difficulty of intercultural communication tells a story about two young fish swimming in the river. An older fish swims by and greets them. Cheerily he adds, "The water is nice and cool today". One young fish turns to the other and asks, "What is water?" Most of the difficulties we encounter with cultures other than our own are due to the fact that our own social norms were handed down to us at an early age and are now mainly unconscious. So they are not immediately obvious to us and we simply take our own cultural norms for granted. Our understanding of another culture is necessarily filtered through our own subjective cultural viewpoint. This means that we cannot avoid bias when we make a judgement or generalisation about another culture. It is necessarily subjective. In order to begin to discuss cultural differences we need an objective way of describing them, one that offers a practical way forward.

The problem was largely solved by Geert Hofstede in the 1970s when he was working with the large multi-national IBM Corporation. Hofstede sent out an extensive questionnaire concerning values and attitudes to 117,000 IBM employees distributed around 50 countries. Factor analysis of the returns allowed Hofstede to identify six scales which he called cultural dimensions. Each cultural dimension is a numerical scale that represents varying attitudes to some aspect of social organisation and values. The *Theory of*

Cultural Dimensions proposes that any national culture can be represented by a set of six indices, each index representing a relative position within each one of the six cultural dimensions (Hofstede et al., 2010). The power of Hofstede's theory lies in providing us with a way of comparing the relative positions of societies as if it were from a detached viewpoint outside the framework as a whole. More recently Hofstede's theory has been validated by independently collected data sets and investigations.

Hofstede's six cultural dimensions are:

1   ***Power distance*** – the degree to which cultures are either egalitarian or hierarchical.
2   ***Individualism vs collectivism*** – the relative importance of the group compared with individuals
3   ***Masculinity vs. femininity*** – the degree to which gender roles are distinct or shared
4   ***Uncertainty avoidance*** – accepting rules contrasted with tolerance for ambiguity
5   ***Long term vs. short term orientation*** – the relative importance of planning for the future
6   ***Indulgence vs restraint*** – the degree to which gratification of natural desires is tolerated.

In a similar approach to cultural dimensions, Erin Meyer directed her attention to organisational management styles that are typical of various national cultures (Meyer, 2014). Based on a study of a large number of international students at the INSEAD Business School in France, Meyer describes eight scales that characterise organisation cultures in different countries. Some overlap with Hofstede's cultural dimensions and some to not:

Communicating – clarity and directness *vs.* sophisticated with nuanced layers of implied meaning (also called low *vs.* high context)
Evaluating – frankness and direct criticism *vs.* diplomacy and softer expression
Persuading –begins by explaining principles *vs.* begins with facts and examples
Leading –egalitarian *vs.* hierarchical *(i.e.* Power Distance)
Deciding –top-down *vs.* consensual management styles *(i.e.* individualism *vs.* collectivism)

Trusting – task-based *vs.* relationship-based management (similar
to masculinity *vs.* femininity)
Disagreeing – confrontational *vs.* non-confrontational styles
Scheduling –promptness and logical order *vs.* addressing new opportunities as they arise.

The attempts by Hofstede, Meyer and others to find an objective
way of describing cultures are not without their critics. The positions on the various cultural dimensions taken by any one nation
are those of an idealised 'typical' person from that culture. Obviously, there must be a broad spread about the average so that the
tails of each statistical distribution overlap with those of other cultures. The United States, for example, is strongly dominated by an
individualist culture, although some minority groups exhibit collectivist cultural styles (*e.g.* First Nations, Mormons and Shakers).
It does not necessarily follow that any individual mentee must hold
the same views as those of most of his/her countrymen. Despite
these criticisms, the Theory of Cultural Dimensions does allow ob-
jectively measurable and replicable generalisations. It can be made
use of in a consistent and detached way.

Perhaps the greatest benefit that this approach confers is to re-
move subjective notions of superiority or inferiority of any one culture compared to any other. All societies have evolved in response
to their local environment and to the history of their region in a way
that optimises the survival benefits to their people. Therefore every
culture is due respect for success within its own context. It is clear
that a basic knowledge of the different cultural dimensions allows us
to better understand the likely responses of our mentees and to be
more flexible in our own responses. *Effective cross-cultural mentoring relies on self-awareness, effective self-management, and flexibility.*

Given the number of cultural dimensions already described,
there is a huge range of possible intercultural mentoring pairings.
Detailed and specific advice on intercultural mentoring is not feasible in the space available here. An approach that works well in
one country may fail entirely in another. However, some general
observations may be helpful for readers from Anglophone nations
for mentoring nationals of other countries in which they travel:

- *Active listening* is the single most important thing you can do.
- *Understanding the Theory of Cultural Dimensions* can be a great
  help. Learn as much as you can about the culture and language

before and during your visit. The more you know and the more experience you have, the better will be your mentoring. In many countries it will be is taken as a sign of your respect and interest if you learn even a few words of the local language.

- *Advice on behaviour* from nationals of the country you visit who have themselves been expatriates in your country can be very helpful. Similarly, listening to long term expatriates in the country you visit can be very helpful, especially those who have local marriage partners and families. However, look out for unconscious biases in expatriates generally.

- *In high power distance countries*, you will be treated as having relatively high status and you may feel embarrassed by what can seem like unwarranted deference. Despite that, good manners are always important. You must know about the local forms of respect and what you might be inclined to do or say that could give offence. In Arab countries, for example, be careful to avoid eating or handing things over with your left hand. Junior professionals will often be psychologically comfortable interacting with you in what is likely to strike you as a subservient manner. If you try to insist on equality you may cause your mentees considerable discomfort. You can really only help them when they learn to trust you. At that point conversations around how to express independent opinions while maintaining a respectful demeanour may be helpful. The unexamined assumption that professionals in the same field have been trained in the same way as ourselves and so think in the same way often turns out to be wrong.

- *In collectivist countries*, group loyalty and coherence are to be expected. Your apparent partiality to one person, even if it is unintentional, may be wrongly taken as favouring one whole group over another. Demanding that people take individual responsibility goes against cultural norms and requires creative handling. There is often a strong emphasis on what is taken to be "correct". Again, encouraging independent thinking alongside respect for the whole group can be helpful in mentoring.

- *In high context countries* a great deal remains unsaid and much will be taken as implied, often in error. Pay close attention to non-verbal communication. Scoping and contracting are therefore especially important. Be careful to clarify mutual expectations. For delicate matters it may be considered impolite to raise your concerns directly, so make sure that you build trust and that meetings are private. In this, confidentiality is especially important. In this regard non-confrontational countries are often similar to those with high context and high power distance.

CASE HISTORY 5.4 – A SUCCESSFUL MENTORING ENGAGEMENT IN WHICH A MENTOR IS BROUGHT DOWN TO EARTH (GUS HANCOCK)

In a 35 year academic career I have mentored ("supervised" within my research group) some 45 graduate students who have obtained doctorates, and some 100 final year undergraduate project students. Most have been British, but one from a southeast Asian country gave me cause for thinking about my own mentoring ability. The project was carefully planned, much discussion was held about the strategy, time lines and procedures, and very soon some excellent experimental results emerged. They were presented to me with pride. After going through an exhaustive procedure of checking and validation which showed to my satisfaction that they were correct I asked the student "Now, what do you think is the explanation for this chemical behaviour?" The student looked puzzled. "I've taken the measurements. You are the professor. Your job is to explain them to me". I realised that there were cultural differences between the roles of mentor and mentee that I did not appreciate. I persisted. "No", I said, "I would like you to tell me what you think. You've already read a lot of past literature in this area, you have a very good grasp of the theory, and I'd like to hear your views. But remember, this is a new area of research, and there may not at present be a "correct" answer. But take a few days and let me know your thoughts." The student looked very anxious and I made sure that other more experienced members of the group could offer help and support if needed.

It wasn't. A few days later a perfectly logical explanation was presented to me, I found out that it had been through the student's own efforts, and our relationship became closer to mentor and mentee than the supervisor/student relationship that had been assumed. I felt pleased that my mentoring abilities had produced success. The student was invited to present the work at a prestigious international conference – the only graduate student to do so in company of senior post-docs and professors. The presentation (which had been meticulously prepared) was received with much applause, and, on the front row of the lecture theatre, I basked in the reflected glory of my achievements. The chairman called for questions, and I expected that they would be gentle and quite straightforward for a graduate student to answer. The graduate student smiled. "I would like you all to be aware that this is my very first presentation at such a meeting, and I am delighted that as the only graduate student speaker I have been given the opportunity to address you. But I do think that my supervisor ought to have a chance to say something, and I am sure that he would be delighted to answer any questions". The lecture theatre erupted in laughter and applause, and I got a total and deserved grilling from the audience. The graduate student produced an excellent thesis and is now a distinguished professor at a European University.

Chapter 6

Mentor training and organisational mentoring schemes

6.1 MENTOR TRAINING AND SUPERVISION

Having come this far, I hope it is now clear that there's more to mentoring than just giving some good-natured advice. Perhaps you'll think there's a lot more, too much to remember and put it into practice in the flurry of a mentoring session. Reading, although it is an essential part of learning, is not enough by itself. In fact, all of mentoring requires dual processing (as described in Section 3.3) and that means that one should achieve the skill level of unconscious competence by practicing the skills and trying out the ideas under observation.

At the time of writing there are relatively few accredited mentor training courses available in Europe or America. There are, however, many excellent executive coach training courses run by reputable and well-established training organisations. Although coaching is distinct from mentoring, most of the skills, principles and processes are directly transferrable (see Section 2.3). The significant benefit of taking a coaching or mentoring training course is to have your practice observed by experts during the training programme. As Bandura's Social Learning Theory has shown, that is the best way to learn any skill (see Section 5.3).

When choosing a training programme, it will be important to select one that is accredited by one of the major coaching institutions – the European Mentoring and Coaching Council (EMCC), the International Coach Federation (ICF) and the Association for Coaching (AC). Individually and in collaboration, the coaching and mentoring institutions regulate accreditation not only for practitioners but also for training courses. Their websites give an abundance of information on what is covered in accredited courses of various levels as well as names and contact details of approved training providers. Hawkins and Smith (2006) give details of coach and mentor training core principles and development that can be used to assess the content of training programmes.

The term 'supervision' is used by mentors, coaches and coaching institutions in the same sense as it is used in counselling and psychotherapy. A supervisor in this sense is someone knowledgeable and very experienced who helps individual coaches and mentors with their practice and especially with specific problems, naturally using his/her own coaching skills. Coaching institutions require that accredited coaches maintain a specific level of supervision., A supervisor is effectively a mentor's mentor. All mentors in science, technology and engineering would benefit from having a supervisor.

6.2 AN OUTLINE FOR AN ORGANISATIONAL MENTORING SCHEME

The benefits of good mentoring are so obviously desirable that many major organisations have set up their own in-house mentoring schemes. As we have seen, there are well-authenticated reports of dysfunctional mentoring and many organisational pitfalls. As Clutterbuck has pointed out, an estimated 40% of mentoring schemes fail in some or all of their stated aims (Clutterbuck, 2011). So, what would an effective mentoring scheme look like?

Adapting Clutterbuck's statement, to be effective any mentoring scheme would achieve:

- A clear organisational purpose (*e.g.* improving retention of mentees by 25% or more)
- Most mentees' personal development objectives
- Learning by most of the mentors
- Willingness of both parties to engage in mentoring (as mentor or mentee) again.

Megginson *et al.* (2006) give a series of case studies of organisational mentoring, the design of, and lessons from which are applicable generally. What follows here is a very basic outline intended for readers who would like to set up a scheme to get started. It would, however, be advisable for any organisation considering setting up a mentoring scheme to retain an expert who specialises in coaching and mentoring training to help with its development.

The first essential element in creating any mentoring scheme is organisational 'buy-in'. It is normal to find that interest and enthusiasm for a new scheme is relatively easy to stimulate, if only because everyone knows that the benefits of good mentoring are considerable. Later on after some contact with reality, minds might begin to change and enthusiasm often wanes. The kinds of problems that dysfunctional mentoring throws up are described above (see Section 2.4). Administration problems can add substantially to individual failures. Senior staff may worry about the added burden of mentoring on the small amount of time they have to spare from their primary responsibilities. Many mentors doubt their own capability in mentoring, concerned about how far their mentoring duties should extend and the degree of responsibility that they carry for their mentee's progress. Anxieties on the mentees' part include doubts about what they might be asked to do and fear of the risk of exposing their professional weaknesses to their employers. Cynics whisper that the scheme is yet another well-intentioned but misguided company initiative and just tick the boxes. In the end the bright new mentoring scheme can drop below the horizon where all those previous company experiments have gone.

Some organisations will find it easier than others to set up a mentoring scheme depending on whether or not the principles of mentoring conflict with the existing organisational culture. For example, some traditional heavy engineering companies have long-established command-and-control hierarchical conservative cultures which can come into conflict with the principles of Client-centred and Intrinsic Motivation. The most effective mentoring schemes may be found in those organisations where staff supervision and recruitment policies support individual initiative and self-responsible professional attitudes and where top level management gives full support.

'Buy-in' for the scheme is achieved primarily through the agency and energy of someone with authority and influence who champions and promotes the idea for setting up the mentoring scheme.

He/she gathers a small band of allies who start an informal pilot scheme. As the scheme is shown to work and as initial benefits begin to appear, so the promoters gradually include more and more people and build momentum. The most effective mentoring scheme will be based on:

- Mentor training by an accredited coach training organisation (Section 6.1)
- Adherence to the principles of mentoring (Section 4.1)
- Structuring mentoring engagements (Section 4.2)
- Management support at the highest level.

The Self-responsibility Principle requires mentors to accept responsibility for the quality of their work as mentors. This means that they need training as mentors and must also undertake continuous professional development (CPD), not only in their own professional subject area but also in the art of mentoring. Training should be delivered by one of the well-established coaching and mentoring training organisations that are accredited by professional coaching institutions.

The scheme must be voluntary for all participants. If mentees and mentors sense any coercion or pressure then failure of the whole scheme is almost guaranteed. The early movers will be intrinsically motivated. Organisational support is required to allow a planned and agreed amount of time away from ordinary duties while trusting the mentoring pairs to use their time wisely. The mentors' voluntary contributions of their effort and time should attract no more reward than credit for CPD. Similarly, mentees should understand that they must take responsibility for their own professional development in mentoring, thereby also gaining CPD credit. In matching mentees with mentors a simple mechanism is required that allows both mentors and mentees to decline to engage with the other without any retribution or questions asked.

The Ethical Responsibility Principle and the need to establish trust require that the mentees' confidentiality is maintained. In some organisations this requirement can give difficulty for line managers. A mechanism is required for the organisation to be reassured that the time and effort spent in mentoring is worthwhile. The best answer is to be found in the executive coaching practice of three-way meetings. The mentee, his/her line manager and the mentor hold an initial three-way meeting to discuss the mentee's development needs, the organisational support available and what independent evidence of progress would be acceptable to all three parties. So, for example, if a young scientist wishes to develop her public speaking ability, she contacts a mentor who agrees to hold a scoping meeting and then to help. Her line manager naturally wants to know how she is progressing. Together they agree that the mentee's progress will be evaluated after she has read a paper at a specified conference in the future. A second three-way meeting will be held after the conference to evaluate progress. Neither mentor nor mentee will be required to tell the line manager confidential details of anything that happened during the mentoring sessions.

In cases where line managers are also mentors for their own staff, the Client-centred Mentoring Principle can come into conflict with organisational requirements. As a general rule, mentoring pairs should not be selected from the same department. Exceptions can be made although both parties should be aware of the potential for a conflict of interest. A possible exception might be a short period of mentoring where the line manager is obviously the most qualified person available to offer help on a specific technical topic.

A mentoring scheme requires not only a structured approach to individual mentoring but also careful design of the scheme as a whole. Although every scheme must be custom-built to serve the unique development needs of the mentees and the organisation in question, most schemes will have some features in common. A suggested cycle of events for each mentoring engagement could be:

1   *Professional development review* – the line manager and each candidate mentee agree and plan for the mentee's professional development needs. The organisation must decide on what training and mentoring support it will provide. This is, in any case, a normal part of most organisations' human resources performance management cycle.
2   *Panel of mentors* – Senior scientists and engineers volunteer to sit on a panel of mentors. Prospective mentees are offered access to the panel on a voluntary basis. They can choose from a

list of the individual panel members. Prospective mentees then approach one or more mentors from the panel and together arrange individual scoping meetings.

3   *Scoping meetings* – The prospective mentee and potential mentor meet (as described in Section 4.2). They discuss the purpose and scope of mentoring and, if suitable, agree an informal 'contract' to work together, including dates and times for a limited number of sessions over a fixed period.

4   *Initial three-way meeting* – The line manager, the mentee and the mentor meet together to discuss the aims of the mentoring engagement and an outline schedule. They agree what indicators would be acceptable evidence for the mentee's progress. They also confirm that details of mentoring will be kept confidential.

5   *Mentoring sessions* – the mentoring pair meet for the agreed series of mentoring sessions with the mentee carrying out agreed private study assignments between meetings. A common arrangement would be four 90-minute meetings spaced two to four weeks apart. The series concludes with a review in private between the mentoring pair (see Section 4.2)

6   *Closing three-way meeting* – the line manager, the mentee and the mentor conduct a three-way meeting, as described above. The evidence for progress provided by the previously agreed indicators, is reviewed. Decisions are made on whether or not further mentoring with the same or a different mentor is needed.

Given agreement by all parties, an individual mentee may repeat this cycle of mentoring as often as necessary on the same or different topics with the same or different mentors. The organisation's part in all of this is limited to facilitating the process, making staff resources available to an agreed extent and reviewing the evidence for individual progress. The mentoring scheme can be developed in support of training courses in which the mentors are used to help the trainees embed what they have learned. Examples would be mentoring to support training courses for a new technical direction or development that an organisation plans to make. A graphic that can be used as a basis for discussion about an organizational mentoring scheme is shown in Figure 6.1.

Mentoring Scheme

Scientific & technical leadership responsibilities
- Ensure best practice
- Protect against technical error
- Encourage professional development

Effective mentoring
- Increases professional knowledge & skill
- Embeds formal learning in personal practice
- Enhances careers of both mentees & mentors

Mentor's Responsibility
- Guiding, not primarily instructing
- Facilitating Mentee's thinking
- Encouraging Mentee's professional development
- Increasing Mentee's clarity & self-awareness
- Developing Mentee's communication skills
- Honesty & openness about what you can & can't do
- Building trust and offering support
- Self-management - intuition, emotion, ethics, boundaries
- Giving constructive advice & structured feedback
- Devising & negotiating homework
- Holding Mentee to account (needs prior agreement)
- Managing structure, timing & outcome-orientated focus.

Mentoring Principles
1. **Creating awareness** is the primary purpose
2. **Client-centred** - focusses on the Mentee's needs. Those of sponsoring organisation are secondary
3. **Self-responsibility** - all parties take personal responsibility. All are volunteers. Mentor uses non-directive style. Mentee is responsible for own outcomes.
4. **Intrinsic motivation** - Mentee is naturally self-motivated & capable. Actual performance depends on inhibiting factors. Awareness improves performance
5. **Ethical responsibility** - Mentor has a duty of care. Preserves Mentee's confidentiality. Both parties need openness, honesty & best efforts. No conflict of interest.

Mentee's Responsibility
- Directing own development
- Developing own knowledge & skills - including CPD
- Raising topic for mentoring series & for each session
- Increasing self-awareness
- Developing communication skills
- Making commitments and using best efforts
- Taking responsibility for mentoring outcomes
- Seeking clarity, self-questioning & reflecting
- Negotiating mentoring agreements & homework
- Observing professional ethics
- Respecting Mentor's time & effort

Mentoring Process
Individual discussions to assist Mentee's development
- **Initial Scoping Meeting** - to decide whether or not to work together. Find out what Mentee needs. Mentor explains what can/can't do. Agree schedule & time, typically 3-4 sessions, 60-90 mins each, 2-4 weeks apart
- **Mentoring Sessions** - Build trust in informal, supportive but structured meetings
- Begin with **'Contracting'**- establish Mentee's purpose for each meeting (see below).
- **Mentor raises awareness** by listening and questions. May use a model *e.g.* Goals?-Realities?-Options?-Way forward? Negotiate homework.
- **Review** - at end of each session and end of series

Mentoring Topics
- Tehnical mentoring - knowledge & professional skills
- Career transitions
- Motivational interviewing
- Communication skills
- Interpersonal & management issues
- Self-confidence
- Intercultural mentoring

Mentoring 'Contract'
- Informal mutual agreement about objectives, methods, logistics & expectations.
- Establish contract at start of series & of each meeting.
- What does Mentee want to achieve? What indicators will show progress?
- Agreement to hold to account

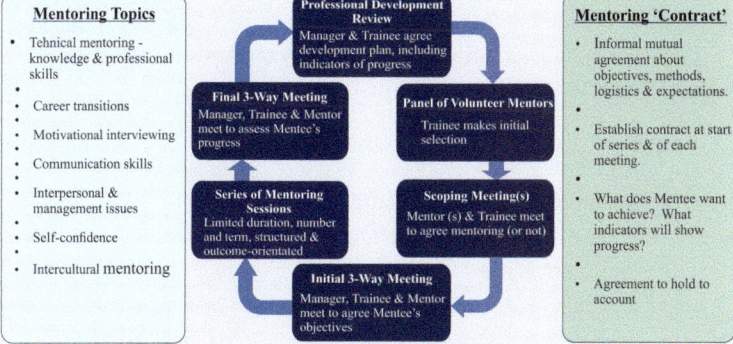

Professional Development Review — Manager & Trainee agree development plan, including indicators of progress

Final 3-Way Meeting — Manager, Trainee & Mentor meet to assess Mentee's progress

Panel of Volunteer Mentors — Trainee makes initial selection

Series of Mentoring Sessions — Limited duration, number and term, structured & outcome-orientated

Scoping Meeting(s) — Mentor (s) & Trainee meet to agree mentoring (or not)

Initial 3-Way Meeting — Manager, Trainee & Mentor meet to agree Mentee's objectives

Figure 6.1 **An outline for a mentoring scheme.**

EXERCISE 6.1 – OBSERVED MENTORING PRACTICE

Total time: 60 minutes. Make up a group of three people (a triad): first person is **Mentor,** second person is **Mentee** and third person is **Observer.** One round takes 15 minutes with a total of three rounds of observed practice plus one round of plenary discussion. **Choose different roles for each round.** Arrange the seating in a triangle with Mentor and Mentee facing each other. The Observer sits apart to one side. Everyone explicitly agrees to keep confidential whatever is revealed.

Round 1: Mentoring session = 10 minutes

Mentor opens by building rapport, contracting and by asking the Mentee to bring an issue for mentoring.

Mentee describes the issue. This is not role-play. It must be a current, genuine, real-world issue.

It may be a personal or professional dilemma, a problem, or something he/she simply wishes to think through.

It can be a technical issue if both are working in the same field. Do **not** to choose a major personal issue with deep emotional loading. Choose something less important, even quite trivial. The subject chosen is simply a vehicle for study of the method and not of interest in its own right.

Examples of minor issues for discussion are:

- a difficult choice for a birthday present
- coping with a bad decision by an adolescent relative or a junior colleague
- a problem about office procedure raised by a colleague
- preparing for a meeting with the boss next week to discuss something unimportant
- negotiating with a relative or friend about choosing an option for entertainment.

The Mentor questions effectively, listens actively and self-manages. Mentor may follow a coaching structure model such as GROW but does not have to. The ideal outcome is an insight which may or may not resolve the presenting issue. Mentor's responsibility is to keep to time. After ten minutes draws the session to a close.

If it over-runs the allocated time, the Observer will terminate the session.

Observed feedback = 5 minutes

Observer takes no part in the mentoring interaction. The Observer's role is to give the Mentor constructive feedback on the method and process *after* the mentoring session. *The purpose is to study the mentoring method and NOT the substance of the Mentee's issue.* Feedback must be *evidence-based.* Observer is looking for how well the essential skills are exercised – skilful questioning, focussed listening, building trust, self-management and giving advice (if appropriate). Also, adherence to principles, contracting and structure. Observer may take some *brief* keyword notes. *Observer to give three positive points and one area for improvement.* Apart from asking for clarification, the triad should *not engage in discussion at this stage.* Then terminate the round.

Round 2 (total 15 minutes)

The triad moves round the triangle and changes places (literally). Repeat: mentoring 10 mins; Feedback 5 mins.

Round 3 (total 15 minutes)

The triad moves round the triangle and changes places (literally). Repeat: mentoring 10 mins; Feedback 5 mins.

Round 4 Plenary Discussion (total 15 minutes) – Q & A:

- *What have you learned about mentoring skills?*
- *What have you learned about structure and procedure?*
- *How did each participant feel in the different roles?*

References

Allen, T.D, Lentz, E., and Day, R. 2006. Career success outcomes associated with mentoring others. A comparison of mentors and non-mentors. *Journal of Career Development.* v. 32, No. 3. pp. 272–285. University of Missouri.

Anderson, L.W. (Ed.), Krathwohl, D.R. (Ed.), Airasian, P.W., Cruikshank, K.A., Mayer, R.E., Pintrich, P.R., Raths, J., and Wittrock, M.C. 2001. *A taxonomy for learning, teaching, and assessing: A revision of Bloom's Taxonomy of Educational Objectives* (Complete edition). Longman, New York.

Association for Coaching. 2020. Website: https://www.association-forcoaching.com/

Bandura, A. 1977. *Social Learning Theory.* Prentice Hall, Englewood Cliffs, NJ.

Beard, C., and Wilson, J.P. 2006. *Experiential Learning: A Handbook for Education, Training and Coaching.* Kogan Page, 3rd Edn, 2013, London, Philadelphia, New Delhi.

Blakey, J., and Day, I. 2012. *Challenging Coaching.* Nicholas Brealey Publishing, London, Boston.

Chartered Institute of Personnel and Development. 2020. *Coaching and mentoring factsheet.* Internet website: https://www.cipd.co.uk/knowledge/fundamentals/people/development/coaching-mentoring-factsheet

Clutterbuck, D. 2011. *Why mentoring programmes and relationships fail.* Online article, Website Clutterbuck Associates accessed 23 October 2020: www.clutterbuckassociates.com

Covey, S. 1989. *The Seven Habits of Highly Effective People.* Simon & Schuster, London.

Dobelli, R. 2013. *The Art of Thinking Clearly.* Sceptre Books, Hodder & Stoughton, London.

Drucker, P.F. 2005. Managing oneself. In The best of HBR 1999. *Harvard Business Review.*

Dweck, C.S. 2012. *Mind-Set. How You Can Fulfil Your Potential.* Constable & Robinson Ltd., London.

Eby, L.T., McManus, S.E., Simon, S.A., and Russell, J.E.A. (2000). The protégé's perspective regarding negative mentoring experiences: The development of a taxonomy. *Journal of Vocational Behaviour,* v. 57, pp. 1–21.

European Mentoring & Coaching Council. 2020. Online article *EMCC Competence Framework V2.* Website accessed 23 October 2020: https://www.emccglobal.org/.

Fowler, H.H., and Fowler, F.G. (Editors –based on the Oxford Dictionary). 1964. *Concise Oxford Dictionary of Current English.* Oxford University Press, London, 5th Edn.

Gallwey, W.T. 2002. *The Inner Game of Work.* TEXERE Publishing Ltd., London.

Garvin, D.A., and Margolis, J.D. 2015. The art of giving and receiving advice. *Harvard Business Review.* Jan–Feb, pp. 61–70.

Goleman, D. 1996. *Emotional Intelligence.* Bloomsbury Publishing, London.

Goleman, D. 2006. *Social Intelligence.* Arrow Books, London.

Guirdham, M. 2002. *Interactive Behaviour at Work.* Person Education Ltd., Harlow, 3rd Edn.

Hargie, O. (Ed.). 2006. *The Handbook of Communication Skills.* Routledge, London and New York, 3rd Edn.

Hargie, O., and Dickson, D. 2004. *Skilled Interpersonal Communication.* Routledge, London and New York, 4th Edn.

Hawkins, P., and Smith, N. 2006. *Coaching, Mentoring and Organisational Consultancy.* Open University Press, McGraw-Hill Education, Maidenhead.

Hofstede, G., Hofstede, G.J., and Minkov, M. 2010. *Cultures and Organizations. Software of the Mind. Intercultural Cooperation and Its Important for Survival.* McGraw-Hill, New York, 3rd Edn.

Holland, C. 2009. Workplace mentoring: A literature review. *Work and Education Research & Development Services,* New Zealand Gov.

Honey, P., and Mumford, A. 1992. *Manual of Learning Styles.* Honey Publications, Maidenhead, 3rd Edn.

International Coach Federation. 2020. Website: https://coachfederation.org/

Krathwohl, D.R., Autumn 2002. A revision of Bloom's taxonomy, an overview. *Theory into Practice,* v. 41, No. 4, pp. 212–225. College of Education, The Ohio State University.

Leary-Joyce, J. 2014. *The Fertile Void: Gestalt Coaching at Work.* Academy of Executive Coaching Press, St. Albans.

Megginson, D., Clutterbuck, D., Garvey, B., Stokes, P., and Garrett-Harris, R. 2006. *Mentoring in Action.* Kogan Page, London, Philadelphia, 2nd Edn.

Mental Health Foundation. 2020. Website accessed 6 October 2020: https://www.mentalhealth.org.uk/statistics/mental-health-statistics-most-common-mental-health-problems

Meyer, E. 2014. *The Culture Map. Decoding How People Think, Lead and Get Things Done across Cultures.* Public Affairs, New York.

Murphy, W., and Kram, K.E. 2014. *Strategic Relationships at Work.* McGraw-Hill Education, New York.

Nauta, M.M. 2010. The development, evolution, and status of Holland's theory of vocational personalities: Reflections and future directions for counselling psychology. *Journal of Counselling Psychology*, v. 57, No. 1, pp. 11–22. American Psychological Association.

O*NET Interest Profiler. Sponsored by US Department of Labor, Employment & Training Administration. Website - https://www.mynextmove.org/explore/ip. Site updated April 20, 2021.

Olivero, G., Bane, K.D., and Kopelman, R.E. 1997. Executive coaching as a transfer of training tool: Effects on productivity in a public agency. *Public Personnel Management.* v. 26, No. 4, pp. 461–69. Winter issue.

Passmore, J., and Whybrow, A. 2008. Motivational interviewing. A specific approach for coaching psychologists. In Palmer, S., and Whybrow, A. (Editors) *Handbook of Coaching Psychology*, Chap. 3, pp. 160–173. Routledge. London and New York.

Patterson, K., Grenny, J., Maxfield, D., McMillan, R., and Switzler, A. 2013. *Crucial Accountability. Tools for Resolving Violated Expectations, Broken Commitments and Bad Behaviour.* McGraw-Hill Education, New York, 2nd Edn.

Seligman, M. 2011. *Flourish. A New Understanding of Happiness and Well-Being.* Nicolas Brealey Publishing. London and Boston.

Starr, J. 2003. *The Coaching Manual.* Pearson Education Ltd, London, 2nd Edn.

Starr, J. 2014. *The Mentoring Manual.* Pearson Education Ltd, London.

Tredennick, H. 1959. *Plato: The Last Days of Socrates.* Penguin Books, London and Tonbridge.

Whitmore, J. 2002. *Coaching for Performance: Growing People, Performance and Purpose.* Nicholas Brealey. London, Naperville, USA.

Index

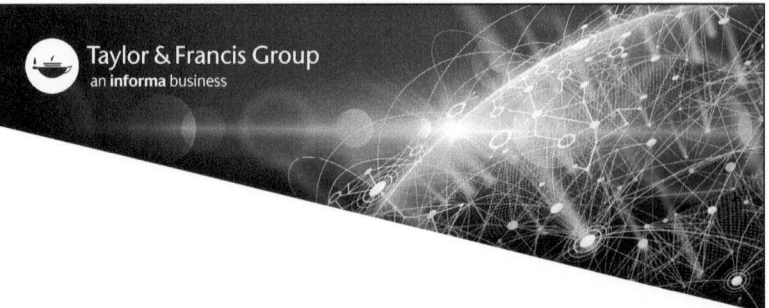